T0207605

Springer Undergraduate Mathematics Series

The Springer Undergraduate Mathematics Series (SUMS) is a series designed for undergraduates in mathematics and the sciences worldwide. From core foundational material to final year topics, SUMS books take a fresh and modern approach. Textual explanations are supported by a wealth of examples, problems and fully-worked solutions, with particular attention paid to universal areas of difficulty. These practical and concise texts are designed for a one- or two-semester course but the self-study approach makes them ideal for independent use.

Robert Magnus

Essential Ordinary Differential Equations

Robert Magnus
Mathematics Division
University of Iceland
Reykjavik, Iceland

ISSN 1615-2085 ISSN 2197-4144 (electronic)
Springer Undergraduate Mathematics Series
ISBN 978-3-031-11530-1 ISBN 978-3-031-11531-8 (eBook)
https://doi.org/10.1007/978-3-031-11531-8

Mathematics Subject Classification: 34-01, 34A12, 34A30, 34B09, 34B24, 34B27, 34B05

This Springer imprint is published by the registered company Springer Nature Switzerland AG
The registered company address is: Gewerbestrasse 11, 6330 Cham, Switzerland

To the memory of
Albert and Gertrude Magnus

Preface

The title of this book indicates a personal view concerning what topics should be included under the heading of "essential." I am sure that many readers will disagree with some of my choices. Is it really essential to include so much existence theory? Is Sturm-Liouville theory essential? Should not some numerical methods be included? And what about some qualitative studies of plane systems?

What guided me was a simple thought. Many undergraduate students of mathematics will study differential equations to a limited extent only. Which topics could be viewed as necessary knowledge to serve up in a course for undergraduates, bearing in mind that differential equations pervade the whole of physics and engineering, as well as many allied branches of mathematics? We should also bear in mind that those wishing to learn more will doubtless be offered a variety of optional courses, going beyond the essential topics, and tailored to the specific interests of the student. Some will include theoretical studies of dynamical systems; others will include computational mathematics. While other students will have no interest in pursuing differential equations any further, but should surely come away knowing something useful.

The topics presented here could form the basis of such a course. Most of the material should be accessible to a student in their second (or certainly third) undergraduate year. Throughout the text we have mathematics students in mind. Therefore, a high standard of rigour is maintained where possible. Moreover, essential does not mean simple or elementary. The topics selected are treated in some depth. Demands are made of the reader from the start. In the opening paragraph, we plunge straight into the theory of linear differential equations of order n, without tiptoeing through the second order case first. From Chap. 5 onwards, we probably go a bit beyond the essential and the material becomes in parts quite challenging, but any teacher will exercise choice over what to omit.

There is absolutely no reason why a text of this kind should try to be self-contained, and develop from scratch all methods and concepts used. No course exists in its own vacuum. It is assumed that the reader is acquainted with multivariate calculus, for example partial derivatives, the chain rule, multiple integrals, and understands concepts of rigorous analysis, such as limits, function series and uniform convergence. Linear algebra (vector spaces and matrices) is used without hesitation. Occasionally complex analysis is needed, and is especially relied on in Chap. 7. The Lebesgue integral is avoided, which causes some awkwardness, again, especially in Chap. 7. A little metric space theory is occasionally needed, but mainly things that the reader may have acquired in a course of multivariate calculus, such as the Bolzano-Weierstrass theorem and the notion of a compact set in $\mathbb{R}^n$. In Chap. 7, it could help if the reader understands that there are different convergence notions for function series. Again, in connection with Chap. 7, it would be nice if the student was concurrently learning some classical Fourier series. Broadly speaking, we rely on the mathematics that the student has learned in their first year and is currently learning in their second or third year. Chapters 6 and 7 make the greatest demands on the reader's knowledge of analysis and metric spaces, whilst Chaps. 5 and 6 make the greatest demands for knowledge of linear algebra.

A large number of exercises is included, some with hints. Many are quite challenging. There are also sections presenting so-called "projects," (for want of a better term). In them, text and exercises alternate; the latter are enumerated by Arabic numerals prefixed by a letter "A," "B," etc., to make for easy reference. The projects go a bit beyond the essential, and if this work is used as a teaching text, the new ideas that are introduced can be studied by students on their own, or in groups, and be the basis of student presentations. A list of all the projects can be perused by consulting "Projects" in the index at the back of this text.

Reykjavik, Iceland
April 2022

Robert Magnus

Contents

Chapter 1
Linear Ordinary Differential Equations

> Can't bring back time. Like holding water in your hand.

A linear, nth order (where n is a positive integer), ordinary differential equation in standard form is the problem

$$y^{(n)} + p_{n-1}(x)y^{(n-1)} + \cdots + p_1(x)y' + p_0(x)y = g(x), \tag{1.1}$$

where the coefficient functions[1] $p_0(x), \ldots, p_{n-1}(x)$ and the function $g(x)$ are defined and real valued in an open interval I of the real line $\mathbb{R}$. The interval I can be bounded or unbounded.

The function g is called the inhomogeneous term. If $g = 0$ (that is g is the zero function taking the value 0 for all values of its argument x) then we say that the problem is homogeneous. Otherwise it is called non-homogeneous or inhomogeneous.

A solution of (1.1) is an n-times differentiable function $y(x)$ defined in the interval I, such that

$$\frac{d^n y}{dx^n}(x) + p_{n-1}(x)\frac{d^{n-1}y}{dx^{n-1}}(x) + \cdots + p_1(x)\frac{dy}{dx}(x) + p_0(x)y(x) = g(x) \tag{1.2}$$

for all x in I.

The different components of (1.1) have conventional names. The variable x is called the independent variable. The variable y is called the dependent variable; we

[1] We do not aim for consistency in the way we talk about functions, sometimes following the custom of practical calculus to speak of "the function $f(x)$". Otherwise we speak of "the function f", especially if ambiguity may arise, for example if we wish to assert that f is zero.

R. Magnus, *Essential Ordinary Differential Equations*, Springer Undergraduate Mathematics Series, https://doi.org/10.1007/978-3-031-11531-8_1

seek to know how y depends on x, as a function $y(x)$. The functions $p_0, \ldots, p_{n-1}$ are called the coefficients. The function g is called the inhomogeneous term.

1.1 First Order Linear Equations

The first order linear equation is

$$y' + p(x)y = g(x) \tag{1.3}$$

By a *general solution* of a differential equation is meant a function of x involving arbitrary constants $A, B, \ldots$ which is a solution for all values of the constants and is such that all possible solutions are obtained by assigning appropriate values to the constants.

Obviously a general solution is not unique (just as a vector space does not have a unique basis). Nevertheless, we often use the definite article and speak of "the general solution", somehow regarding them as being equivalent to each other. Sometimes a general solution is referred to as a *complete solution*.

Proposition 1.1 *If p and g are continuous in the interval I then a general solution of* (1.3) *is given by*

$$y(x) = e^{-M(x)}\left(C + \int g(x)e^{M(x)}\,dx\right) \tag{1.4}$$

where C is an arbitrary constant, $M(x) := \int p(x)\,dx$ is an antiderivative of $p(x)$ in I and $\int g(x)e^{M(x)}\,dx$ is an antiderivative of $g(x)e^{M(x)}$ in I.

Notes

(1) A formula such as (1.4) that solves a differential equation by means of antiderivatives is traditionally called a solution by quadrature, that being an old name for integration. The first order linear equation requires in general two quadratures.

(2) A function f continuous in an interval I possesses an antiderivative in I. It is a solution of the simplest possible differential equation $dy/dx = f(x)$. The fundamental theorem of calculus provides one of the form

$$F(x) = \int_{x_0}^{x} f(t)\,dt, \quad (x \in I)$$

where x_0 is an arbitrarily chosen base point in I (but there may be others not obtainable by selecting a base point x_0).

(3) If F is an antiderivative of f then so is $F + C$ where C is any constant function. Any two antiderivatives of F in the same interval differ by a constant. Altering

the choice of the two antiderivatives in (1.4) gives no new solution because of the already arbitrary constant C, a point that the reader should check.

Proof That formula (1.4) is a solution for all values of the constant C is easy to check (another thing for the reader to do). Now let $y(x)$ be a solution. Then for $x \in I$ we have

$$y'(x) + p(x)y(x) = g(x),$$

and so

$$e^{M(x)}y'(x) + p(x)e^{M(x)}y(x) = g(x)e^{M(x)}.$$

But the left-hand side of this is the derivative of $e^{M(x)}y(x)$ and so

$$e^{M(x)}y(x) = C + \int g(x)e^{M(x)}\,dx,$$

which gives (1.4) and ends the proof. □

1.2 The *n*th Order Linear Equation

No method is known that solves the general nth order, linear differential equation by quadratures if $n \geq 2$. Only special cases, such as equations with constant coefficients, can be solved by these methods.

Functions on an interval I can be added pointwise to build new functions

$$(f_1 + f_2)(x) = f_1(x) + f_2(x), \quad (x \in I).$$

They can be multiplied by scalars

$$(\lambda f)(x) = \lambda f(x), \quad (x \in I).$$

They therefore form a vector space over the real number field $\mathbb{R}$. All references to vector spaces, for the time being, will mean vector spaces over the field $\mathbb{R}$.

The following proposition is now easy to check (and the reader should do it):

Proposition 1.2 *The solutions of an nth order, homogeneous, linear differential equation*

$$y^{(n)} + p_{n-1}(x)y^{(n-1)} + \cdots + p_1(x)y' + p_0(x)y = 0 \qquad (1.5)$$

constitute a vector space of functions on the interval I.

Of course the zero function $y = 0$ is a solution of (1.5) but *a priori* we do not know whether there are others if $n \geq 2$ (the case $n = 1$ being satisfactorily dealt with by Proposition 1.1). We now state the fundamental existence theorem for (1.5). We shall prove it in a later chapter.

Proposition 1.3 *Let the coefficient functions $p_0, \ldots, p_{n-1}$ be continuous in the open interval I. Let $x_0 \in I$ and let numbers $a_1, \ldots, a_n$ be given. Then the problem* (1.5) *has a unique solution in the interval I that satisfies the n conditions*

$$y(x_0) = a_1, \quad y'(x_1) = a_2, \quad \ldots \quad y^{(n-1)}(x_0) = a_n \tag{1.6}$$

Problem (1.5) together with the conditions (1.6) is called the *initial value problem*, or *Cauchy problem*, for the differential equation (1.5). The conditions (1.6) are called the *initial conditions* or *Cauchy conditions*. From Proposition 1.3 we now deduce a remarkable result:

Proposition 1.4 *Let the coefficient functions $p_0, \ldots, p_{n-1}$ be continuous in the open interval I. Then the space of solutions of* (1.5) *is an n-dimensional vector space of functions defined in the interval I.*

Proof Let the solution space of the homogeneous equation be the vector space E of n-times differentiable functions on the interval I. Fixing x_0 in I we consider the map $\kappa : E \to \mathbb{R}^n$ given by

$$\kappa(y) = (y(x_0), y'(x_0), \ldots y^{(n-1)}(x_0))$$

The coordinates of the vector $(y(x_0), y'(x_0), \ldots y^{(n-1)}(x_0))$ are called the Cauchy data of the solution $y(x)$. Proposition 1.3 says that κ is surjective (since there exists a solution for arbitrarily given Cauchy data) and injective (since a solution with given Cauchy data is unique). Since κ is obviously linear it is a linear isomorphism between vector spaces. Hence E is n-dimensional. □

We often encounter linear equations for which the leading coefficient (that is the coefficient of $y^{(n)}$) is not equal to 1. Such an equation has the form

$$p_n(x)y^{(n)} + p_{n-1}(x)y^{(n-1)} + \cdots + p_1(x)y' + p_0(x)y = 0$$

Before applying Proposition 1.4 one should divide through by $p_n(x)$ to reduce the equation to standard form. This is only possible in an interval in which $p_n(x)$ has no zeros. Studying the effect of zeros of $p_n(x)$ will take up a later chapter of this text.

Calculating a solution basis for the equation (1.5) is a major problem of practical importance which will occupy a substantial part of this text.

1.2.1 The Wronskian

Let $u_1, \ldots, u_n$ be a sequence of n functions each of which is $n-1$ times differentiable in an interval I. We do not assume that they are the solutions of any differential equation. The determinant

$$W(u_1, \ldots, u_n)(x) := \begin{vmatrix} u_1(x) & \ldots & u_n(x) \\ u_1'(x) & \ldots & u_n'(x) \\ \vdots & & \vdots \\ u_1^{(n-1)}(x) & \ldots & u_n^{(n-1)}(x) \end{vmatrix}$$

is called the Wronskian of the functions $u_1, \ldots, u_n$. It is a function on the interval I.

Proposition 1.5 *Let the coefficient functions $p_0, \ldots, p_{n-1}$ be continuous in the open interval I. Let $u_1, \ldots, u_n$ be n solutions of (1.5) on the interval I and let $x_0 \in I$. Then the solutions $u_1, \ldots, u_n$ form a basis for the space E of all solutions of (1.5) on I if and only if $W(u_1, \ldots, u_n)(x_0) \neq 0$.*

Proof In the proof of Proposition 1.4 we saw that the mapping $\kappa : E \to \mathbb{R}^n$, that maps each solution to its vector of Cauchy data at x_0, is a vector space isomorphism. It follows that n solutions $u_1, \ldots, u_n$ form a basis of E if and only if the n vectors $\kappa(u_1), \ldots, \kappa(u_n)$ form a basis of $\mathbb{R}^n$. The necessary and sufficient condition for this is that the determinant $W(u_1, \ldots, u_n)(x_0)$ is not 0, since these vectors constitute its columns. □

Observing that the point x_0 in Proposition 1.5 can be any point in the interval I, we obtain the following:

Proposition 1.6 *Let the coefficient functions $p_0, \ldots, p_{n-1}$ be continuous in the open interval I. Let $u_1, \ldots, u_n$ be n solutions of (1.5) on the interval I. Then either the Wronskian $W(u_1, \ldots, u_n)(x)$ is non-zero for every x in I or it is zero for every x in I.*

Even though Proposition 1.6 follows at once from Proposition 1.5 there is an interesting formula due to Abel that leads to the same conclusion. It shows that $W(u_1, \ldots, u_n)(x)$ satisfies a first order linear differential equation.

Proposition 1.7 *Let the coefficient functions $p_0, \ldots, p_{n-1}$ be continuous in the open interval I. Let $x_0 \in I$. Let $u_1, \ldots, u_n$ be n solutions of (1.5) on the interval I. Then*

$$W(u_1, \ldots, u_n)(x) = W(u_1, \ldots, u_n)(x_0)e^{-\int_{x_0}^{x} p_{n-1}(t)\,dt}, \quad (x \in I). \tag{1.7}$$

Proof The conclusion follows if we show that $W(x) := W(u_1, \ldots, u_n)(x)$ satisfies the first order differential equation

$$W' = -p_{n-1}(x)W$$

We shall write out the proof for the case $n = 2$ and leave the general case to the reader. All that is involved is the differentiation of a determinant whose entries are functions of x. Since an $n \times n$ determinant is a multilinear function of its n rows, this is done by differentiating each row separately and adding the resulting n determinants. For the case of the Wronskian all the resulting determinants except the last one are zero. One repeatedly uses the fact that a determinant with two equal rows is zero.

In the case $n = 2$ (which serves as a good guide to the general case) we have:

$$
\begin{aligned}
W'(x) &= \frac{d}{dx}\begin{vmatrix} u_1(x) & u_2(x) \\ u_1'(x) & u_2'(x) \end{vmatrix} \\
&= \begin{vmatrix} u_1'(x) & u_2'(x) \\ u_1'(x) & u_2'(x) \end{vmatrix} + \begin{vmatrix} u_1(x) & u_2(x) \\ u_1''(x) & u_2''(x) \end{vmatrix} \\
&= \begin{vmatrix} u_1(x) & u_2(x) \\ u_1''(x) & u_2''(x) \end{vmatrix}
\end{aligned}
$$

Using the differential equation $y'' + p_1(x)y' + p_0(x)y = 0$ satisfied by u_1 and u_2 now gives

$$
\begin{aligned}
W'(x) &= \begin{vmatrix} u_1(x) & u_2(x) \\ -p_1(x)u_1'(x) - p_0(x)u_1(x) & -p_1(x)u_2'(x) - p_0(x)u_2(x) \end{vmatrix} \\
&= -p_1(x)\begin{vmatrix} u_1(x) & u_2(x) \\ u_1'(x) & u_2'(x) \end{vmatrix} - p_0(x)\begin{vmatrix} u_1(x) & u_2(x) \\ u_1(x) & u_2(x) \end{vmatrix} \\
&= -p_1(x)W(x)
\end{aligned}
$$

as claimed for the case $n = 2$. □

1.2.2 Non-homogeneous Equations

Proposition 1.8 *Let the coefficient functions* $p_0, \ldots, p_{n-1}$ *be continuous in the open interval* I *and let the function* g *be continuous in* I. *Let* $v(x)$ *be a solution on* I *of the non-homogeneous equation*

$$
y^{(n)} + p_{n-1}(x)y^{(n-1)} + \cdots + p_1(x)y' + p_0(x)y = g(x). \tag{1.8}
$$

Let $u_1, \ldots, u_n$ be a solution basis for the homogeneous equation. Then the general solution of (1.8) *can be written*

$$y(x) = c_1 u_1(x) + \cdots + c_n u_n(x) + v(x) \tag{1.9}$$

where $c_1, \ldots, c_n$ are arbitrary constants.

The function $v(x)$ in this proposition is traditionally called a *particular solution* of the non-homogeneous equation. In spite of its name, any solution can be used as a particular solution.

Proof In the first place the formula (1.9) is a solution of (1.8) however the constants $c_1, \ldots, c_n$ are chosen. Now let $y(x)$ be a solution of (1.8). Consider the function $z(x) := y(x) - v(x)$. It is easily seen that $z(x)$ satisfies the homogeneous equation, and hence is of the form $c_1 u_1(x) + \cdots + c_n u_n(x)$ for a certain choice of the constants. □

Next we prove the existence of a particular solution. In fact we do more. We show that once a solution basis is known for the homogeneous equation a particular solution can be found by quadratures. The classical procedure is called the method of *variation of parameters.*

Proposition 1.9 *Let the coefficient functions $p_0, \ldots, p_{n-1}$ be continuous in the open interval I and let the function g be continuous in I. Let the functions u_1, $\ldots$, u_n form a solution basis for the homogeneous equation. Consider the linear combination*

$$v(x) = c_1(x)u_1(x) + \cdots + c_n(x)u_n(x)$$

where the coefficients $c_1(x), \ldots, c_n(x)$ are functions instead of constants.[2] *A sufficient condition for $v(x)$ to be a solution of* (1.8) *is that the functions $c_1(x)$, $\ldots$, $c_n(x)$ satisfy the matrix equation*

$$\begin{bmatrix} u_1(x) & \ldots & u_n(x) \\ u_1'(x) & \ldots & u_n'(x) \\ \vdots & & \vdots \\ u_1^{(n-1)}(x) & \ldots & u_n^{(n-1)}(x) \end{bmatrix} \begin{bmatrix} c_1'(x) \\ c_2'(x) \\ \vdots \\ c_n'(x) \end{bmatrix} = \begin{bmatrix} 0 \\ 0 \\ \vdots \\ g(x) \end{bmatrix} \tag{1.10}$$

Note The non-vanishing of the Wronskian ensures that the matrix equation can be solved for the vector $(c_1'(x), \ldots, c_n'(x))$ and then the coefficient functions $c_1(x)$, $\ldots$, $c_n(x)$ are found by integration.

[2] Hence the name: *variation of constants* or *parameters.*

Proof Differentiate the formula for $v(x)$ repeatedly and use the successive lines of the matrix relation (1.10). We obtain (including the formula for $v(x)$ placed initially)

$$\begin{array}{rclclcl} v(x) & = & c_1(x)u_1(x) & +\cdots+ & c_n(x)u_n(x) & + & 0 \\ v'(x) & = & c_1(x)u_1'(x) & +\cdots+ & c_n(x)u_n'(x) & + & 0 \\ v''(x) & = & c_1(x)u_1''(x) & +\cdots+ & c_n(x)u_n''(x) & + & 0 \\ \vdots & & \vdots & & \vdots & & \vdots \\ v^{(n-1)}(x) & = & c_1(x)u_1^{(n-1)}(x) & +\cdots+ & c_n(x)u_n^{(n-1)}(x) & + & 0 \\ v^{(n)}(x) & = & c_1(x)u_1^{(n)}(x) & +\cdots+ & c_n(x)u_n^{(n)}(x) & + & g(x) \end{array}$$

Now we multiply the first line by $p_0(x)$, the second by $p_1(x)$ and so on, down to the last line which is multiplied by 1. The lines are then added together. Since $u_1, \ldots,$ u_n all satisfy the homogeneous equation, the total of each column on the right-hand side except the last is zero, leaving $g(x)$ standing alone. The result is

$$v^{(n)}(x) + p_{n-1}(x)v^{(n-1)}(x) + \cdots + p_1(x)v'(x) + p_0(x)v(x) = g(x)$$

□

The second order case

$$y'' + p_1(x)y' + p_0(x)y = g(x)$$

leads to the problem

$$\begin{bmatrix} u_1(x) & u_2(x) \\ u_1'(x) & u_2'(x) \end{bmatrix} \begin{bmatrix} c_1'(x) \\ c_2'(x) \end{bmatrix} = \begin{bmatrix} 0 \\ g(x) \end{bmatrix}$$

The solution can be written by Kramer's rule as

$$c_1'(x) = -W(x)^{-1}u_2(x)g(x), \quad c_2'(x) = W(x)^{-1}u_1(x)g(x)$$

where $W(x)$ is the Wronskian of u_1, u_2. This gives the particular solution

$$v(x) = -u_1(x)\int W(x)^{-1}u_2(x)g(x)\,dx + u_2(x)\int W(x)^{-1}u_1(x)g(x)\,dx$$

the integrals here being antiderivatives. This formula is often useful, especially when Proposition 1.7 is used to calculate $W(x)^{-1}$.

Proposition 1.10 *Let the coefficient functions $p_0, \ldots, p_{n-1}$ be continuous in the open interval I. Let the function g be continuous in I, let $a_1, \ldots, a_n$ be given numbers and let $x_0 \in I$. Then the* initial value problem

$$y^{(n)} + p_{n-1}(x)y^{(n-1)} + \cdots + p_1(x)y' + p_0(x)y = g(x) \tag{1.11}$$

$$y(x_0) = a_1, \quad \ldots, \quad y^{(n-1)}(x_0) = a_n$$

has a unique solution in I.

Proof By Proposition 1.8 Eq. (1.11) has a general solution

$$y(x) = c_1u_1(x) + \cdots + c_nu_n(x) + v(x)$$

where $u_1, \ldots, u_n$ is a solution basis for the homogeneous equation and $v(x)$ is a particular solution of (1.11). Substituting the initial values we find that the coefficients c_k must satisfy

$$\begin{aligned}
c_1u_1(x_0) + \cdots + c_nu_n(x_0) &= a_1 - v(x_0)\\
c_1u_1'(x_0) + \cdots + c_nu_n'(x_0) &= a_2 - v'(x_0)\\
\vdots \qquad\qquad & \quad\vdots \qquad \vdots\\
c_1u_1^{(n-1)}(x_0) + \cdots + c_nu_n^{(n-1)}(x_0) &= a_n - v^{(n-1)}(x_0)
\end{aligned}$$

By the non-vanishing of the Wronskian of the solutions $u_1, \ldots, u_n$, we see that these equations determine the coefficients c_k uniquely. □

In this proposition we see the immense importance of the problem of finding a solution basis for the homogeneous equation. First steps towards solving this problem in any general sort of way will be taken in Sect. 1.3.

1.2.3 Complex Solutions

One often encounters complex valued solutions of linear differential equations, and such solutions may be useful even though the equation itself has real valued coefficients and does not seem to invite a passage to the complex realm. For example, the equation $y^{(4)}+y = 0$ has a solution basis comprising the four functions

$$e^{x/\sqrt{2}}\cos(x/\sqrt{2}), \quad e^{x/\sqrt{2}}\sin(x/\sqrt{2}), \quad e^{-x/\sqrt{2}}\cos(x/\sqrt{2}), \quad e^{-x/\sqrt{2}}\sin(x/\sqrt{2}),$$

but if we allow complex valued solutions we can write another basis, algebraically simpler, comprising the functions

$$e^{\omega x}, \quad e^{i\omega x}, \quad e^{-\omega x}, \quad e^{-i\omega x},$$

where $\omega = (1+i)/\sqrt{2}$. The second basis might prove useful even when we seek, at the end of the day, a real valued solution. How these bases can be found is a topic taken later in this chapter.

For now we note that complex valued functions defined in the real interval I form a vector space over the complex field $\mathbb{C}$. They can be differentiated in the obvious way, by differentiating the real part and the imaginary part, thus:

$$y'(x) = u'(x) + iv'(x),$$

where $y(x) = u(x) + iv(x)$ and the functions u and v are real valued. It is now obvious that $u + iv$ is a solution of (1.5) (which has only real valued coefficient functions) if and only if u and v are individually real valued solutions. This says that the space of complex valued solutions is the *complexification* of the space of real solutions; it is a vector space over $\mathbb{C}$ with dimension n.

We can go further and suppose that the coefficient functions $p_1, \ldots, p_n$ have complex values, as well as the inhomogeneous term g and the initial values. The analogues of the propositions of this section hold for complex equations without change, although they do not obviously follow from Proposition 1.3. The vector space of solutions of the homogeneous equation will be an n-dimensional vector space over $\mathbb{C}$ of complex valued functions. However, in this chapter we restrict ourselves to equations with real coefficients, as their properties can be derived from the as yet unproved Proposition 1.3.

It is important to understand that the independent variable x is always real. At this point, we do not need differentiation with respect to a complex variable, which leads to the theory of *complex analytic functions*. The notion of a *differential equation in the complex domain*, for which a solution is a function of a complex variable, is not touched upon in this text.

1.2.4 Exercises

1. Find a general solution for the following equations:

   (a) $y' - y = x^2$

   (b) $y' - \dfrac{1}{x}y = \dfrac{1}{x^2}, \quad (x > 0)$

   (c) $y' - \dfrac{1}{x^2}y = \dfrac{1}{x^3}, \quad (x > 0)$

   (d) $y' - (\tan x)y = \cos x, \quad (-\pi/2 < x < \pi/2)$.

Note In items (b), (c) and (d) it is essential to specify an interval, because of the blowing up of the coefficient functions or the inhomogeneous term.

(e) For the equations in items (a) and (d), find the unique solution that satisfies $y(0) = 0$.
(f) For the equations in items (b) and (c), find the unique solution that satisfies $y(1) = 0$.

The formula (1.4) solves the first order linear equation by integrals, but the integrals are not necessarily elementary.

2. The error function is defined by

$$\mathrm{erf}\,(x) = \frac{2}{\sqrt{\pi}} \int_0^x e^{-t^2}\, dt, \quad (-\infty < x < \infty).$$

Express the solution of the Cauchy problem

$$y' - 2xy = 1, \quad y(1) = 0$$

in terms of the error function.

Some non-linear equations can be transformed into linear equations. This is illustrated in the next exercise, and again in Exercise 16.

3. The non-linear equation

$$y' + p(x)y = g(x)y^{m+1}$$

is called a Bernoulli equation. The number m can be any real number except 0. Let $y = u^{-1/m}$. Show that u satisfies a first order linear equation. Use this and Proposition 1.1 to solve the Bernoulli equation

$$y' - y = xy^2.$$

Sketch the graphs of some solutions.
Note It is likely that you obtain a form of the general solution, containing a single constant C, covering all solutions whose graphs lie in the part of the plane for which $y \neq 0$, but no value of C will give the solution $y = 0$. This is not surprising, since the change of variables by means of which one passes from y to u is not defined at $y = 0$. You may also observe that the interval in which a solution is defined varies with the solution. This is quite different from what one sees with linear equations.
4. The equation $y'' + y = 0$ plays a basic role in the theory of the circular functions and is doubtless well known to the reader. Show that the functions $A\cos x + B\sin x$ and $C\cos x + D\sin x$ form a solution basis if and only if $AD - BC \neq 0$.

5. Find a general solution in terms of integrals of the equation

$$y'' + y = \tan x$$

on the interval $]-\frac{\pi}{2}, \frac{\pi}{2}[$. Carry out the integrations.

6. Show that the functions $u_1(x) = x$ and $u_2(x) = x \ln x$ form a basis for the space of all solutions of

$$y'' - \frac{1}{x}y' + \frac{1}{x^2}y = 0$$

on the interval $]0, \infty[$. What would you use as a basis for the space of solutions on $]-\infty, 0[$?

7. Let $p(x)$ be a continuous function in the open interval I, let $x_0 \in I$, and let u_1 and u_2 be solutions of

$$y'' + p(x)y = 0$$

on I that satisfy the conditions

$$u_1(x_0) = 1, \quad u_1'(x_0) = 0, \quad u_2(x_0) = 0, \quad u_2'(x_0) = 1.$$

Show that the Wronskian $W(u_1, u_2)(x)$ of u_1, u_2 satisfies

$$W(u_1, u_2)(x) = 1$$

for all $x \in I$.

8. (a) Let u_1, u_2 be differentiable functions in the open interval I. Suppose that u_1 and u_2 are linearly dependent as functions on I. Show that their Wronskian $W(u_1, u_2)(x)$ is 0 for all x in I.
   Note It is not assumed that u_1 and u_2 are solutions of a linear differential equation $y'' + p_1(x)y' + p_0(x)y = 0$; so this is not a consequence of Proposition 1.5.
   Hint Suppose that $au_1(x) + bu_2(x) = 0$ for all x in I where a and b are not both 0, and differentiate.
   (b) Prove the same result for n functions (skipping, if you so wish, the case of two functions), assuming that they have derivatives to order $n - 1$.

Without additional assumptions (such as that u_1 and u_2 are solutions of some linear differential equation) the converse of the previous exercise is false, as we see in the next exercise.

9. Let $u_1(x) = x^2$, $u_2(x) = x|x|$. Show that both functions are differentiable for all x (including $x = 0$), their Wronskian is 0 for all x, but they are not linearly dependent on any open interval that contains 0.

So if the Wronskian is 0 in I, what can we conclude about the functions if we can't conclude that the functions are linearly dependent on I?

10. Let the Wronskian $W(u_1, u_2)(x)$ of the differentiable functions u_1 and u_2 be 0 for all x in I. Let x_0 be a point in I such that $u_1(x_0) \neq 0$. Show that there exists an open interval $J \subset I$ containing x_0, such that the restrictions of u_1 and u_2 to J are linearly dependent. If you care to, you can frame a similar result for n functions.

Care must be taken when studying equations for which the leading coefficient is not 1. The next two exercises illustrate this.

11. Consider the functions $u_1(x) = x^3$ and $u_2(x) = x^4$.
    (a) Check that they both satisfy the equation
    $$x^2y'' - 6xy' + 12y = 0.$$
    (b) Compute their Wronskian and observe where it is 0. Explain why this does not contradict Proposition 1.6.
    (c) Show that the function
    $$v(x) = \begin{cases} Ax^3 + Bx^4, & \text{if } x < 0 \\ Cx^3 + Dx^4, & \text{if } x \geq 0 \end{cases}$$
    is twice continuously differentiable and satisfies the equation of item (a) in the interval $]-\infty, \infty[$ for all choices of the constants A, B, C and D. It would appear that the solution space is at least four-dimensional. Why does this not contradict Proposition 1.4?

12. Find the general solution of
    $$xy' - y = x + 1$$
    (a) on the interval $]-\infty, 0[$;
    (b) on the interval $]0, \infty[$.

    What happens if you try to solve the Cauchy problem for the equation with the condition $y(0) = a$?
13. There is no procedure for solving the general second order equation
    $$y'' + p(x)y' + q(x)y = 0 \tag{1.12}$$
    using integrals. But if we already know one solution $u(x)$ that is nowhere 0 in an interval I, then the complete solution on I can be obtained by integrals. The procedure is called *reduction of order* and reduces the problem to solving a first order linear equation.

Let $y = u(x)v$ where $u(x)$ is a known solution and v a new dependent variable. Show that v satisfies

$$v'' + \left(\frac{2u'(x)}{u(x)} + p(x)\right)v' = 0$$

and use formula (1.4) to obtain the general solution of (1.12) in the form

$$y(x) = Au(x)\int \frac{e^{-P(x)}}{u(x)^2}\,dx + Bu(x)$$

where $P(x)$ is an antiderivative for $p(x)$ and A and B are arbitrary constants. *Note* You can use the same substitution to reduce the inhomogeneous equation

$$y'' + p(x)y' + q(x)y = g(x)$$

to a first order equation for v', and hence solve it, circumventing variation of parameters.

14. Obtain the general solution of $y'' + xy' - y = 0$ in view of the fact that x is a solution.
15. Find a general solution in terms of integrals of the equation

$$y'' - \frac{1}{x}y' + \frac{1}{x^2}y = g(x)$$

on the interval $]0, \infty[$ given that $g(x)$ is continuous in that interval (a solution basis for the homogeneous equation was given in Exercise 6). Carry out the integrations for the case $g(x) = \ln x$.

16. The non-linear equation

$$y' + p(x)y + q(x)y^2 = r(x)$$

is called a Riccati equation.

(a) Suppose that in the interval I the functions $p(x)$ and $r(x)$ are continuous, the function $q(x)$ is continuously differentiable and $q(x)$ is nowhere 0. Let $y = u'/q(x)u$ where u is a new dependent variable. Show that u satisfies the second order linear equation

$$u'' + \left(p(x) - \frac{q'(x)}{q(x)}\right)u' - r(x)q(x)u = 0$$

A solution $u(x)$ of this equation gives a solution $y = u'(x)/q(x)u(x)$ of the Riccati equation on any interval in which $u(x)$ is nowhere 0.

(b) Using the method of the previous item, transform the equation

$$y' = x^2 + y^2$$

into a linear equation.

17. Let u_1 and u_2 form a solution basis for the homogeneous equation

$$y'' + p_1(x)y' + p_0(x)y = 0, \quad (a < x < b)$$

and let $W(x) := W(u_1, u_2)(x)$ be their Wronskian. Let $a < x_0 < b$. Show that the solution to the initial value problem

$$y'' + p_1(x)y' + p_0(x)y = g(x), \quad y(x_0) = y'(x_0) = 0$$

is given by

$$y(x) = -u_1(x) \int_{x_0}^{x} W^{-1} u_2 g + u_2(x) \int_{x_0}^{x} W^{-1} u_1 g.$$

1.2.5 *Projects*

A. *Project on separation theorems*

We consider a second order homogeneous equation

$$p_2(x)y'' + p_1(x)y' + p_0(x)y = 0 \tag{1.13}$$

where p_0, p_1, p_2 are continuous and real valued in the open interval I and $p_2(x) \neq 0$ for all $x \in I$.

A1. Let $y(x)$ be a real valued solution of (1.13) that is not identically zero. Show that $y(x)$ can have only finitely many zeros in any bounded and closed subinterval of I.
Hint If $y(x)$ has infinitely many zeros in an interval $[c, d] \subset I$ then, by the Bolzano-Weierstrass theorem, those zeros will have a limit point in $[c, d]$.

The result states that the zeros of $y(x)$ form a discrete set, necessarily closed, and so it makes sense, given a zero x_1 of $y(x)$, to talk of the next zero above x_1 or the next zero below x_1.

A2. Let $u(x)$ and $v(x)$ be two real linearly independent solutions of (1.13). Let $x_1 < x_2$ be consecutive zeros of u. Show that $v(x)$ has exactly one zero in $[x_1, x_2]$ and that it lies in the open interval $]x_1, x_2[$.

Hint Suppose that $v(x)$ is non-zero in the interval $[x_1, x_2]$ and consider the function $u(x)/v(x)$.

This result is called a separation theorem: the zeros of linearly independent solutions *separate each other*. It applies to real solutions but not to complex valued solutions.

A3. Show that if $AD - BC \neq 0$ then the zeros of

$$u(x) = A\cos x + B\sin x, \qquad v(x) = C\cos x + D\sin x$$

separate each other.
Hint They are solutions of $y'' + y = 0$. However, the complex solutions e^{ix} and e^{-ix} of the same equation have no zeros.

B. *Project on comparison theorems*

We compare two equations on the same interval I, written in the *Sturm-Liouville form*

$$\frac{d}{dx}\left(K(x)\frac{dy}{dx}\right) - G_1(x)y = 0 \tag{1.14}$$

$$\frac{d}{dx}\left(K(x)\frac{dy}{dx}\right) - G_2(x)y = 0 \tag{1.15}$$

where $K(x)$ is continuously differentiable in the interval I, $K(x) > 0$ in I and $G_1(x)$ and $G_2(x)$ continuous in I. All the coefficient functions are assumed to be real valued. We shall have a lot more to say about problems cast in this form in Chap. 7.

B1. Assume that $G_1(x) > G_2(x)$ for each $x \in I$. Let u be a solution of (1.14), let v a solution of (1.15), and let $x_1 < x_2$ be consecutive zeros of u. Prove that v has a zero in the open interval $]x_1, x_2[$.
Note The function v can be zero at x_1 or x_2, but we explicitly ask for a zero strictly between x_1 and x_2.
Hint We may suppose that $u(x) > 0$ in the open interval $]x_1, x_2[$. Assuming that $v(x) > 0$ in the open interval $]x_1, x_2[$, consider the integral

$$\int_{x_1}^{x_2} \big(G_1(x) - G_2(x)\big)u(x)v(x)\,dx.$$

Examine what (1.14) and (1.15) reveal about its sign and obtain a contradiction.

B2. Consider the problem

$$y'' + p(x)y = 0. \tag{1.16}$$

Suppose that $p(x)$ is continuous in $]-\infty, \infty[$ and satisfies $p(x) > m$ for all x, where m is a positive constant. Show that all solutions of (1.16) have infinitely many zeros and that the distance between consecutive zeros of a non-identically-zero solution is strictly less than $\pi/\sqrt{m}$.

C. *Project on Green's function: a foretaste*

We study the inhomogeneous nth order equation

$$y^{(n)} + p_{n-1}(x)y^{(n-1)} + \cdots + p_1(x)y' + p_0(x)y = g(x), \qquad (x \in I)$$

where the coefficient functions and the inhomogeneous term g are continuous in the interval I. The aim is to develop a formula analogous to that obtained in Exercise 17. The outcome is Green's function for the initial value problem. The notion of Green's function is more usually associated with *boundary value problems*, in which context its characteristic properties appear more clearly. This is a topic taken up in detail later in this chapter.

Let $x_0 \in I$ and let $u_1, \ldots, u_n$ be a basis for the solution space of the homogeneous equation.

C1. Choose the antiderivatives in the variation of parameters method so that they satisfy $c_1(x_0) = \cdots = c_n(x_0) = 0$. Show that the corresponding particular solution $v(x)$ satisfies

$$v(x_0) = v'(x_0) = \cdots = v^{(n-1)}(x_0) = 0.$$

C2. For each $\xi \in I$ let $G(x, \xi)$ be the solution of the homogeneous equation that satisfies the Cauchy conditions

$$y(\xi) = 0, \quad y'(\xi) = 0, \ \ldots, \ y^{(n-2)}(\xi) = 0, \quad y^{(n-1)}(\xi) = 1.$$

Letting the functions $c_1(x),\ldots, c_n(x)$ and $v(x)$ be as in the previous item, show that

$$G(x, \xi)g(\xi) = c_1'(\xi)u_1(x) + \cdots + c_n'(\xi)u_n(x), \quad (x,\ \xi \in I)$$

and deduce the formula

$$v(x) = \int_{x_0}^{x} G(x, \xi)g(\xi)\, d\xi.$$

1.3 Homogeneous Linear Equations with Constant Coefficients

In this section we study the homogeneous equation

$$p_n y^{(n)} + p_{n-1} y^{(n-1)} + \cdots + p_1 y' + p_0 y = 0 \tag{1.17}$$

with constant coefficients $p_0, \ldots, p_n$. We assume here that $p_n \neq 0$ so we could (but do not) divide throughout by p_n to convert the equation to standard form. The solution space is an n-dimensional vector space of functions on the real line $]-\infty, \infty[$. If the coefficients are real and we admit only real valued functions then it is n-dimensional over $\mathbb{R}$. If we admit complex valued solutions (and we are forced to do this if some coefficients are not real) then it is n-dimensional over $\mathbb{C}$. In this chapter we only study equations with real coefficients, but it may be still be advantageous to allow complex valued solutions.

Closely associated with the differential equation is the polynomial

$$P(X) := p_n X^n + p_{n-1} X^{n-1} + \cdots + p_1 X + p_0$$

and the so-called *indicial equation*[3]

$$P(X) = 0.$$

The roots of the indicial equation play a fundamental role in the theory of the linear equation with constant coefficients.

Proposition 1.11

1. *The function $e^{\lambda x}$ is a solution of* (1.17) *if and only if λ is a root of the indicial equation.*
2. *Suppose that the indicial equation has n distinct roots $\lambda_1, \ldots, \lambda_n$, possibly complex. Then the functions*

$$e^{\lambda_1 x}, \ldots, e^{\lambda_n x}$$

form a solution basis.

Proof

1. Left to the reader.
2. They are solutions by item 1. Moreover they are linearly independent. This can be seen (for example) from the Wronskian computed at $x = 0$. This is the Vandermonde determinant

---

[3] Also called the characteristic equation. The polynomial $P(X)$ is similarly often called the characteristic polynomial.

$$\begin{vmatrix} 1 & \dots & 1 \\ \lambda_1 & \dots & \lambda_n \\ \vdots & & \vdots \\ \lambda_1^{n-1} & \dots & \lambda_n^{n-1} \end{vmatrix}$$

well known to be non-zero if the numbers $\lambda_1, \dots, \lambda_n$ are distinct (for a suggested proof of this see Exercise 8). □

If the indicial equation has a complex root $\lambda = \alpha + i\beta$ we obtain a complex valued solution

$$e^{\lambda x} = e^{\alpha x}(\cos \beta x + i \sin \beta x).$$

If the coefficients are real numbers then complex roots of the indicial equation occur in conjugate pairs. Thus if $\lambda = \alpha + i\beta$ is a root so is $\bar{\lambda} = \alpha - i\beta$. We have the two complex valued solutions

$$e^{\lambda x} = e^{\alpha x}(\cos \beta x + i \sin \beta x), \quad e^{\bar{\lambda} x} = e^{\alpha x}(\cos \beta x - i \sin \beta x).$$

These span a two-dimensional space of complex valued solutions, but another basis for this same two-dimensional space is the function pair

$$e^{\alpha x} \cos \beta x, \quad e^{\alpha x} \sin \beta x.$$

As these are real valued functions we can use them instead of $e^{\lambda x}$ and $e^{\bar{\lambda} x}$ in a basis if we are only interested in real valued solutions.

Example Find a basis for the solutions of $y^{(4)} - y = 0$.

Solution The indicial equation $\lambda^4 - 1 = 0$ has roots $1, -1, i, -i$. Three examples of a solution basis are:

$$e^x, \ e^{-x}, \ e^{ix}, \ e^{-ix}$$

$$e^x, \ e^{-x}, \ \cos x, \ \sin x$$

$$\cosh x, \ \sinh x, \ \cos x, \ \sin x$$

The last basis is useful because its constituent functions satisfy Cauchy conditions at $x = 0$ with plenty of zeros in them.

1.3.1 What to do About Multiple Roots

If the indicial equation has multiple roots there cannot exist a sufficient number of linearly independent solutions of the form $e^{\lambda x}$ to build a solution basis. For example the equation

$$y'' - 2y' + y = 0$$

has only one solution of the form $e^{\lambda x}$, and that is e^x. Since the solution space here is two-dimensional there must exist a solution linearly independent of e^x, but not of the form $e^{\lambda x}$.

To approach this problem it is useful to introduce the operator D of differentiation. Thus if y is a function, Dy is its derivative y', $D^2 y$ its second derivative y'' and so on. Given a polynomial

$$P(X) = p_n X^n + p_{n-1} X^{n-1} + \cdots + p_1 X + p_0$$

we can replace the indeterminate X by D to produce the *differential operator*

$$P(D) = p_n D^n + p_{n-1} D^{n-1} + \cdots + p_1 D + p_0$$

that acts on a function $y(x)$ by means of the prescription

$$P(D)y(x) = p_n y^{(n)}(x) + p_{n-1} y^{(n-1)}(x) + \cdots + p_1 y'(x) + p_0 y(x).$$

An important property of this conversion of polynomials to differential operators is that it respects sum

$$P(X) + Q(X) \mapsto P(D) + Q(D)$$

and product

$$P(X)Q(X) \mapsto P(D)Q(D).$$

In the language of algebra it is a ring homomorphism from the ring of polynomials in one variable to the ring of differential operators. The verification of these claims is simple and is left to the reader.

Now it is easy to check that

$$P(D)e^{\lambda x} = P(\lambda)e^{\lambda x} \tag{1.18}$$

for all λ (proof left to the reader, see exercise 1). In particular, this implies that $e^{\lambda x}$ is a solution of $P(D)y = 0$ if and only if λ is a root of the indicial equation $P(X) = 0$, a fact we already know.

But it implies more. Differentiate (1.18) with respect to λ. Since differentiation with respect to x and λ commute we find

$$P(D)xe^{\lambda x} = P(\lambda)xe^{\lambda x} + P'(\lambda)e^{\lambda x}.$$

Hence if $P(\lambda_1) = P'(\lambda_1) = 0$, which is to say that λ_1 is a root of $P(X) = 0$ of multiplicity at least 2, then $xe^{\lambda_1 x}$ is a solution.

Taking this further let us differentiate (1.18) k times with respect to λ. The result, by Leibniz's rule, is

$$P(D)(x^k e^{\lambda x}) = \sum_{j=0}^{k} \frac{k!}{j!(k-j)!} P^{(j)}(\lambda)x^{k-j}e^{\lambda x}.$$

Hence if λ_1 is a root of order at least $k+1$ then the function $x^k e^{\lambda_1 x}$ is a solution, since all of $P(\lambda_1)$, $P'(\lambda_1)$, ..., $P^{(k)}(\lambda_1)$ are then 0.

Consider the general case of $P(D)y = 0$ where we do not assume that $P(X)$ has distinct roots. By the *fundamental theorem of algebra* there is a factorisation (unique up to order of factors)

$$P(X) = p_n(X-\lambda_1)^{r_1}(X-\lambda_2)^{r_2}\dots(X-\lambda_m)^{r_m}$$

where $\lambda_1, \dots, \lambda_m$ are the distinct roots of $P(X)$, in general complex numbers, and the root λ_k has multiplicity r_k.

Now we can list some solutions in the following table:

Root	Solutions	Number
λ_1	$e^{\lambda_1 x},\ xe^{\lambda_1 x},\ \dots,\ x^{r_1-1}e^{\lambda_1 x}$	r_1
λ_2	$e^{\lambda_2 x},\ xe^{\lambda_2 x},\ \dots,\ x^{r_2-1}e^{\lambda_2 x}$	r_2
$\vdots$	$\vdots \quad \vdots$	$\vdots$
λ_m	$e^{\lambda_m x},\ xe^{\lambda_m x},\ \dots,\ x^{r_m-1}e^{\lambda_m x}$	r_m

The total number of solutions found is $r_1 + \cdots + r_m = n$, precisely the known dimension of the solution space. Thus if we can show that these functions are linearly independent we can conclude that they constitute a solution basis. The following proof works also in the case of simple roots and circumvents the need to use the Vandermonde determinant.

Proposition 1.12 *The solutions listed in the above table are linearly independent.*

Proof Suppose a linear combination of them gave the zero function. Then we would have

$$f_1(x)e^{\lambda_1 x} + \cdots + f_m(x)e^{\lambda_m x} = 0 \tag{1.19}$$

(as an identity, that is, for all x), where the multipliers $f_1, \ldots, f_m$ are polynomials, at least one of which is not the zero polynomial. We shall show, however, using induction on m, that whenever the identity (1.19) holds for distinct $\lambda_1, \ldots, \lambda_m$ and polynomial multipliers $f_1, \ldots, f_m$ then all the polynomials $f_1, \ldots, f_m$ are zero.

The claim clearly holds if $m = 1$. Suppose that it holds for a given m and assume that an identity

$$f_1(x)e^{\lambda_1 x} + \cdots + f_m(x)e^{\lambda_m x} + f_{m+1}(x)e^{\lambda_{m+1} x} = 0$$

holds, where the exponents $\lambda_1, \ldots, \lambda_{m+1}$ are distinct and the multipliers $f_1, \ldots, f_{m+1}$ are polynomials. We multiply through by $e^{-\lambda_{m+1} x}$. This gives

$$f_1(x)e^{\mu_1 x} + \cdots + f_m(x)e^{\mu_m x} + f_{m+1}(x) = 0$$

where $\mu_k = \lambda_k - \lambda_{m+1}$ $(k = 1, \ldots, m)$. The numbers $\mu_1, \ldots, \mu_m$ are distinct from each other and from 0. Now differentiate repeatedly until the polynomial $f_{m+1}(x)$ is reduced to 0. We obtain an identity

$$g_1(x)e^{\mu_1 x} + \cdots + g_m(x)e^{\mu_m x} = 0$$

with new polynomial multipliers $g_1, \ldots, g_m$. By the induction hypothesis these are all 0. But it is easy to check, using the fact that $\mu_k \neq 0$, that g_k has the same degree as f_k for each k. Hence $f_k = 0$ for $k = 1, \ldots, m$. Finally then, $f_{m+1} = 0$ also. □

1.3.2 *Euler's Equation*

The equation

$$p_n x^n y^{(n)} + p_{n-1} x^{n-1} y^{(n-1)} + \cdots + p_1 x y' + p_0 y = 0, \tag{1.20}$$

where $p_0, \ldots, p_n$ are constants and $p_n \neq 0$, is called Euler's equation of order n. Euler's equation does not have constant coefficients, so that the reader may wonder why it is considered in this section. The answer is that it may be transformed into an equation with constant coefficients, from which the theory of it can be derived most satisfactorily, as will be shown below.

Since the leading coefficient is $p_n x^n$ we must divide by it before applying the existence theorems. As a result we obtain a *singularity* at $x = 0$ and the existence theorems apply to the intervals $]0, \infty[$ and $]-\infty, 0[$, but not to an interval containing 0.

A more general version of Euler's equation is obtained by translating the variable x, thus:

$$p_n(x-x_0)^n y^{(n)} + p_{n-1}(x-x_0)^{n-1} y^{(n-1)} + \cdots + p_1(x-x_0)y' + p_0 y = 0.$$

Now the singularity, the zero of the leading coefficient function $p_n(x-x_0)^n$, is at $x = x_0$.

The coefficients p_k are usually real constants in practice, and we seek real solutions. However, the theory we develop here can also handle complex coefficients.

We limit ourselves to studying (1.20) on the interval $]0, \infty[$. According to the basic existence theorem the solution space is an n-dimensional vector space of functions on this interval.

We make the change of independent variable $t = \ln x$. This implies

$$x^k D_x^k y = D_t(D_t - 1)\dots(D_t - k + 1)y, \quad (k = 0, 1, 2, \dots) \tag{1.21}$$

where D_x denotes the operation of differentiation with respect to x, whilst D_t differentiates with respect to t. The proof of (1.21) is left to the reader (see Exercise 6). The result is a linear differential equation with constant coefficients. It has an indicial equation, quite easy to write down as we only have to replace D_t by X, namely

$$p_n X(X-1)\dots(X-n+1) + p_{n-1}X(X-1)\dots(X-n+2) + \cdots$$
$$\cdots + p_2 X(X-1) + p_1 X + p_0 = 0.$$

Now if λ is a root of the indicial equation we know that $e^{\lambda t}$ is a solution of the transformed equation, so that x^λ is a solution of Euler's equation on the interval $]0, \infty[$. Given that the indicial equation has n distinct roots we obtain the general solution in the form

$$y = A_1 x^{\lambda_1} + \cdots + A_n x^{\lambda_n}.$$

The reader should puzzle out the correct form of a solution basis in the case of multiple roots (see Exercise 7).

A root may be complex. If $\alpha + i\beta$ is a root we have a complex solution $x^{\alpha+i\beta}$. If the coefficients are real and only real solutions are desired we may note that

$$x^{\alpha+i\beta} = x^\alpha(\cos(\beta \ln x) + i\sin(\beta \ln x))$$

and the real and imaginary parts, $x^\alpha \cos(\beta \ln x)$ and $x^\alpha \sin(\beta \ln x))$ are two independent real solutions. They span the same two-dimensional space over $\mathbb{C}$ as the two complex solutions $x^{\alpha \pm i\beta}$.

1.3.3 Exercises

1. Verify formula (1.18).
   Hint Do it first for $P(X) = X^k$.
2. Find a solution basis for the following equations:

   (a) $y^{(3)} + y = 0$
   (b) $y^{(4)} + y = 0$
   (c) $y^{(4)} - 2y'' + y = 0$
   (d) $y^{(4)} + 2y'' + y = 0$

3. Compute

$$(D^5 - D^4 + D^3 - D^2 + D - 1)(D^5 + D^4 + D^3 + D^2 + D + 1)\sin x.$$

   Hint There is a slick way that involves almost no algebra. To help you see it try applying the same operator to the function x, instead of $\sin x$.
4. (a) Show that every solution $y(x)$ of (1.17) satisfies $\lim_{x\to\infty} y(x) = 0$ if and only if every root of the indicial equation has negative real part.
   (b) Suppose that every solution $y(x)$ of (1.17) is bounded on the whole line $\mathbb{R}$. Show that every root of the indicial equation lies on the imaginary axis (that is, has zero real part).
   (c) Show that the converse of item (b) is false. How would you modify the condition on the roots in order to obtain a necessary and sufficient condition for every solution to be bounded?
5. One can go further than Exercise 4(a) (the 'if' part) by estimating the rate of decay as $x \to \infty$. Suppose that all roots of the indicial equation of the problem

$$p_n y^{(n)} + p_{n-1} y^{(n-1)} + \cdots + p_1 y' + p_0 y = 0$$

   have negative real parts. Let the distinct roots be $\lambda_1, \ldots, \lambda_p$ and let $m > 0$ satisfy $\max_k \operatorname{Re} \lambda_k < -m < 0$. Show that for each solution $y(x)$ there exists a constant $C > 0$, such that

$$|y(x)| \le Ce^{-mx} \quad (x \in \mathbb{R}).$$

6. Prove formula (1.21).
   Hint Use induction, noting that the chain rule implies that $D_t = x D_x$.
7. Show that if the indicial equation of (1.20) has a root λ with multiplicity r, then the functions $x^\lambda (\ln x)^k$, for $k = 0, 1, \ldots, r - 1$, are linearly independent solutions.
8. Prove that the Vandermonde determinant (see the proof of proposition 1.11) is non-zero if the numbers $\lambda_1, \ldots, \lambda_n$ are distinct.

Hint Let

$$V_n(\lambda_1, \ldots, \lambda_n) = \begin{vmatrix} 1 & 1 & \ldots & 1 \\ \lambda_1 & \lambda_2 & \ldots & \lambda_n \\ \vdots & & \vdots & \\ \lambda_1^{n-1} & \lambda_2^{n-1} & \ldots & \lambda_n^{n-1} \end{vmatrix}$$

and show by induction that

$$V_n(\lambda_1, \ldots, \lambda_n) = (-1)^{\frac{n(n-1)}{2}} \prod_{1 \le i < j \le n} (\lambda_i - \lambda_j)$$

It helps to observe that $V_n(x, \lambda_2, \ldots, \lambda_n)$ is a polynomial of degree $n - 1$ whose roots are $\lambda_2, \ldots, \lambda_n$.

1.4 Non-homogeneous Equations with Constant Coefficients

We have seen that all solutions of the homogeneous equation with constant coefficients

$$p_n y^{(n)} + p_{n-1} y^{(n-1)} + \cdots + p_1 y' + p_0 y = 0$$

are of the form

$$y(x) = f_1(x)e^{\lambda_1 x} + \cdots + f_m(x)e^{\lambda_m x}$$

where the functions $f_1, \ldots, f_m$ are polynomials with complex coefficients and the exponents $\lambda_1, \ldots, \lambda_m$ are distinct complex numbers.

Definition A function of the form $f_1(x)e^{\lambda_1 x} + \cdots + f_m(x)e^{\lambda_m x}$, where $f_1, \ldots, f_m$ are polynomials with complex coefficients and the exponents $\lambda_1, \ldots, \lambda_m$ are distinct complex numbers, is called an *exponential polynomial.*

The reasoning in the proof of Proposition 1.12 indicates that if $y(x)$ is an exponential polynomial, then its representation as a sum of this kind is unique, up to the order of the summands. Algebraic methods are available for finding a particular solution of the non-homogeneous equation if it has constant coefficients and the inhomogeneous term is an exponential polynomial.

A function of the form $f(x)e^{\lambda x}$ where f is a polynomial of degree p will be called a *pure exponential polynomial* with exponent λ and degree p. Pure exponential polynomials with exponent λ and degree less than m (given that $m \ge 1$) form a vector space over $\mathbb{C}$ of dimension m. Let us denote this space by V_m^λ. To complete the picture we define V_m^λ to consist only of the zero function if $m \leq 0$. The functions

$$e^{\lambda x}, \quad \frac{x}{1!}e^{\lambda x}, \quad \dots \quad \frac{x^{m-1}}{(m-1)!}e^{\lambda x}$$

form a useful basis for V_m^λ.

An important property of V_m^λ is that it is closed under differentiation. If $y \in V_m^\lambda$ then so is Dy. So for each m and λ we have a linear operator $D : V_m^\lambda \to V_m^\lambda$.

Proposition 1.13 *The operator $D : V_m^\lambda \to V_m^\lambda$ has the matrix*

$$\begin{bmatrix} \lambda & 1 & 0 & \dots & 0 \\ 0 & \lambda & 1 & \dots & 0 \\ \vdots & \vdots & \vdots & \ddots & \vdots \\ 0 & 0 & 0 & \dots & \lambda \end{bmatrix}$$

(called a Jordan block matrix; it has λ on the diagonal, 1's on the superdiagonal and 0's everywhere else) with respect to the above basis.

Proof Left as an exercise to the reader. □

Proposition 1.14 *If $\lambda \neq 0$ then $D : V_m^\lambda \to V_m^\lambda$ is a linear bijection. On the other hand $DV_m^0 = V_{m-1}^0$ for $m \geq 1$.*

Proof Left as an exercise to the reader. □

Proposition 1.15 *Let $P(X)$ be a polynomial such that $P(\lambda) \neq 0$. Then $P(D) : V_m^\lambda \to V_m^\lambda$ is a linear bijection.*

Proof The spectrum of the operator $D : V_m^\lambda \to V_m^\lambda$ consists of the single eigenvalue λ. Hence, by the spectral mapping theorem of linear algebra, the spectrum of $P(D) : V_m^\lambda \to V_m^\lambda$ consists of the single eigenvalue $P(\lambda)$. Hence if $P(\lambda) \neq 0$ then $P(D)$ is a bijection.

Instead of using the spectral mapping theorem one can observe that the matrix of $P(D)$ is upper triangular with $P(\lambda)$ on its diagonal. □

This proposition tells us something useful about particular solutions.

Proposition 1.16 *If $P(\lambda) \neq 0$ and $f \in V_m^\lambda$ then the equation $P(D)y = f$ has a unique solution in the space V_m^λ.*

We are going to see how to calculate this solution. But first we need some useful rules.

Proposition 1.17 (The Shift Rule) *If f is a C^∞ function and $P(X)$ a polynomial then*

$$P(D)\left(e^{\lambda x} f(x)\right) = e^{\lambda x} P(D+\lambda) f(x).$$

Proof The formula holds for $P(X) = 1$ and $P(X) = X$ (by Leibniz's rule). Hence it holds for X^k and by taking linear combinations we obtain the rule for all polynomials. □

Proposition 1.18 *Suppose that λ is a root of $P(X) = 0$ with multiplicity r. Then $P(D)V_m^\lambda = V_{m-r}^\lambda$.*

Proof We have $P(X) = Q(X)(X-\lambda)^r$ where $Q(X)$ is a polynomial such that $Q(\lambda) \neq 0$. Let $e^{\lambda x} f(x)$ be an element of V_m^λ, that is, $f(x)$ is a polynomial with degree less than m. Then we find

$$P(D)\left(e^{\lambda x} f(x)\right) = Q(D)(D-\lambda)^r \left(e^{\lambda x} f(x)\right) = Q(D)\left(e^{\lambda x} D^r f(x)\right).$$

We now note that $D^r V_m^0 = V_{m-r}^0$ and $Q(D)V_{m-r}^\lambda = V_{m-r}^\lambda$. □

Proposition 1.19 *If λ is root of $P(X) = 0$ with multiplicity r and if $g \in V_m^\lambda$ then the equation $P(D)y = g(x)$ has a solution in V_{m+r}^λ, but it is not unique.*

The proof should be obvious. The solution is not unique because V_r^λ is a subspace of the space of solutions of the homogeneous equation and any element of it may be added to a particular solution to obtain another particular solution. We can make the solution unique by omitting all terms of the form $ax^j e^{\lambda x}$ with $j < r$.

Finally we can record:

Proposition 1.20 *All solutions of the constant coefficient equation*

$$p_n y^{(n)} + p_{n-1} y^{(n-1)} + \cdots + p_1 y' + p_0 y = g(x)$$

are exponential polynomials if g is an exponential polynomial.

1.4.1 How to Calculate a Particular Solution

We shall describe two purely algebraic methods for finding a particular solution to

$$P(D)y := p_n y^{(n)} + p_{n-1} y^{(n-1)} + \cdots + p_1 y' + p_0 y = g(x)$$

when g is an exponential polynomial.

Firstly we may assume that g is a pure exponential polynomial, for example an element of V_m^λ. For we may write g as a sum of pure exponential polynomials, find a particular solution for each summand and add the results.

The first method is known by variously ridiculous names such as *the method of undetermined coefficients*, *the method of judicious guessing* or *the annihilator method*. We regard the differential equation as a linear problem in the vector space V_m^λ (or V_{m+r}^λ for an appropriate r as shown below) and solve using a basis. The most convenient basis of V_m^λ for calculation is

$$e^{\lambda x}, \quad xe^{\lambda x}, \quad \dots, \quad x^{m-1}e^{\lambda x}.$$

This is equivalent to using an *ansatz* or substitution. We may discern two cases, assuming that $g \in V_m^\lambda$:

(1) *λ is not a root of $P(X)$*. There is a unique particular solution in V_m^λ so we use the substitution

$$y(x) = (a_{m-1}x^{m-1} + a_{m-2}x^{m-2} + \cdots + a_0)e^{\lambda x}$$

and we can determine the coefficients by solving a system of linear algebraic equations.

(2) *λ is a root of $P(X)$ with multiplicity r.* There is a particular solution in V_{m+r}^λ which can be made unique by dropping terms of degree less than r. Therefore we can use the substitution

$$y(x) = x^r(a_{m-1}x^{m-1} + a_{m-2}x^{m-2} + \cdots + a_0)e^{\lambda x}.$$

The second method is usually quicker, does not require us to solve a system of linear algebraic equations, but is harder to explain. It is based on the fact that the matrix of $P(D)$ as an operator in V_m^λ is rather special.

Let's assume that λ is a root of $P(X)$ with multiplicity r and that $f(x)$ is a polynomial of degree less than m. The equation $P(D)y = e^{\lambda x}f(x)$ is known to have a particular solution of the form $e^{\lambda x}g(x)$ where g is a polynomial of degree less than $m + r$. We can include the case when λ is not a root by viewing this assumption as meaning that $r = 0$. By the shift rule we have

$$P(D)e^{\lambda x}g(x) = e^{\lambda x}P(D + \lambda)g(x)$$

so we must find a polynomial $g(x)$ of degree less than $m + r$ that satisfies

$$P(D + \lambda)g(x) = f(x).$$

We can write

$$P(X) = Q(X)(X - \lambda)^r$$

where Q is a polynomial such that $Q(\lambda) \neq 0$. Therefore $g(x)$ satisfies

$$Q(D+\lambda)D^r g(x) = f(x)$$

Now write $R(X) = Q(X+\lambda)$. Since $R(0) \neq 0$ there is a unique polynomial $h(x)$ of degree less than m such that $R(D)h(x) = f(x)$. We can determine $g(x)$ from $h(x)$ by integrating r times (not uniquely if $r \geq 1$). The question is: how do we find $h(x)$? In other words we want the inverse of the operator $R(D) : V_m^0 \to V_m^0$ and this must exist since $R(0) \neq 0$. There is a simple formula for this.

Proposition 1.21 *Suppose that $R(X)$ is a polynomial such that $R(0) \neq 0$. Expand $1/R(X)$ as a power series*

$$\frac{1}{R(X)} = \sum_{k=0}^{\infty} a_k X^k$$

Then the inverse of $R(D) : V_m^0 \to V_m^0$ is the operator $\sum_{k=0}^{m-1} a_k D^k$.

Proof We have

$$\frac{1}{R(X)} = \sum_{k=0}^{\infty} a_k X^k = \sum_{k=0}^{m-1} a_k X^k + \left(\sum_{k=0}^{\infty} a_{k+m} X^k\right) X^m$$

where the series are convergent in some disc $|X| < \rho$. We then find

$$1 = R(X) \sum_{k=0}^{m-1} a_k X^k + X^m S(X) \tag{1.22}$$

where

$$S(X) = R(X)\left(\sum_{k=0}^{\infty} a_{k+m} X^k\right).$$

By these equations $S(X)$ is a power series and $X^m S(X)$ is a polynomial. Hence $S(X)$ is a polynomial and so the operator $S(D)$ makes sense. Replacing X by D in (1.22) and letting I be the identity operator in V_m^0 we find

$$I = R(D)\left(\sum_{k=0}^{m-1} a_k D^k\right) + D^m S(D) = R(D)\left(\sum_{k=0}^{m-1} a_k D^k\right)$$

since $D^m = 0$ in the space V_m^0. □

The same method can be used to calculate the inverse of a matrix $R(A)$ when A is a matrix such that $A^m = 0$ and $R(X)$ a polynomial such that $R(0) \neq 0$. In fact, power series can be avoided. We find polynomials $T(X)$ and $S(X)$ such that

$$1 = R(X)T(X) + X^m S(X).$$

The inverse of $R(A)$ is then $T(A)$. Since $R(X)$ and X^m are coprime polynomials (that is, they have no common polynomial divisor of degree greater than or equal to 1), $S(X)$ and $T(X)$ can be found by the Euclidean algorithm. But it may still be simpler to observe that $T(X)$ is the Maclaurin series of $1/R(X)$ truncated at the $(m-1)$st power of X. And in practice it is enough to know only the series $(1+X)^{-1} = 1 - X + X^2 - X^3 + \cdots$. In the following examples we show how the preceding arguments can be turned into a practical pencil and paper method.

Example Find a particular solution of

$$y'' - y' - y = x^3 e^{-x}.$$

Solution Since the equation is $(D^2 - D - 1)y = x^3 e^{-x}$ we can work out a particular solution as follows. The calculations are justified by Propositions 1.17 (the shift rule) and 1.21.

$$\begin{aligned}
y(x) &= (D^2 - D - 1)^{-1}(x^3 e^{-x}) \\
&= e^{-x}((D-1)^2 - (D-1) - 1)^{-1} x^3 \qquad \text{(by the shift rule)} \\
&= e^{-x}(D^2 - 3D + 1)^{-1} x^3 \\
&= e^{-x}\left(1 - (D^2 - 3D) + (D^2 - 3D)^2 - (D^2 - 3D)^3 + \cdots\right) x^3 \\
&\qquad \text{(using the series for } (1+X)^{-1}) \\
&= e^{-x}\left(1 + 3D + 8D^2 + 21D^3\right) x^3 \quad \text{(truncating at the cubic term)} \\
&= e^{-x}(x^3 + 9x^2 + 48x + 126)
\end{aligned}$$

How to Handle Sines and Cosines

Since $\sin \omega x = \frac{1}{2i} e^{i\omega x} - \frac{1}{2i} e^{-i\omega x}$ and $\cos \omega x = \frac{1}{2} e^{i\omega x} + \frac{1}{2} e^{-i\omega x}$ the equation

$$P(D)y = A \cos \omega x + B \sin \omega x$$

can be solved using the methods of this section, by breaking the right-hand side up into the sum of pure exponential polynomials. However, there are some special tricks that can applied to this equation. Let us assume:

(1) $P(X)$ has real coefficients.
(2) $i\omega$ is not a root of $P(X)$ and $\omega \neq 0$.

Consider the 2-dimensional vector space T_ω spanned by the functions $e^{i\omega x}$ and $e^{-i\omega x}$ over $\mathbb{C}$. An alternative basis is $\cos\omega x$, $\sin\omega x$. Now D maps T_ω to itself. In fact when D is viewed as a linear operator in T_ω it satisfies the relation

$$D^2 = -\omega^2 I$$

where I is the identity map in T_ω. Let now $g \in T_\omega$ and view the problem

$$P(D)y = g$$

as a linear problem in the 2-dimensional space T_ω. Now we may replace all occurrences of D^2 in $P(D)$ by $-\omega^2$. We obtain a problem of the form

$$(aD + b)y = g$$

Exercise Prove that a and b are not both 0.

As an example of this reduction we can reduce

$$(D^{10} - D^7 + D^4 - 1)y = \cos 2x$$

to

$$\big((-4)^5 - (-4)^3 D + (-4)^2 - 1\big)y = \cos 2x,$$

and simplifying we obtain

$$(64D - 1009)y = \cos 2x.$$

We now solve the reduced equation, either by using the usual method for a first order equation, or else by applying the following result, the proof of which is left to the reader:

Proposition 1.22 *Let a and b be real numbers not both 0, and let $\omega \neq 0$. Then the operator*

$$aD + b : T_\omega \to T_\omega$$

is invertible. Its inverse is the operator

$$-\frac{a}{\omega^2 a^2 + b^2} D + \frac{b}{\omega^2 a^2 + b^2}.$$

Another method which is applicable if $P(X)$ has real coefficients is to replace the equation

$$P(D)y = A\cos\omega x$$

by

$$P(D)y = Ae^{i\omega x},$$

solve, and take the real part of the solution. Similarly,

$$P(D)y = B\sin\omega x$$

is replaced by

$$P(D)y = Be^{i\omega x},$$

and the imaginary part of the solution taken. This method also works for the resonant cases (that is to say, those for which $i\omega$ is a root of the indicial equation), and for the exponential polynomial cases

$$P(D)y = f(x)\sin\omega x, \qquad P(D)y = f(x)\cos\omega x$$

where $f(x)$ is a polynomial.

1.4.2 Exercises

1. Find a general solution of the following equations:

   (a) $y'' - 4y = x^2 e^x$
   (b) $y'' - 4y = x^2 e^{2x}$
   (c) $y'' + 4y = x^2 \cos x$
   (d) $y'' + 4y = x^2 \cos 2x$
   (e) $y'' + y = \cos^3 x$

2. Find a 2π-periodic solution of the equation

$$y'''' + y = \cos x, \quad (-\infty < x < \infty)$$

   and prove that it is the only bounded solution.
   Hint To prove uniqueness show that the homogeneous equation has no bounded solution except $y = 0$.

3. We mentioned the name “annihilator method” in connection with finding a particular solution to $P(D)y = g(x)$ when $g(x)$ is an exponential polynomial. In fact the said method is a way to find a suitable Ansatz for a solution. Suppose that $P(X)$ has degree n. The method consists in the following steps and the exercise for the reader is to check that they lead to the desired result:

   (a) Find a polynomial $Q(X)$ such that $Q(D)g = 0$ (an annihilator). Suppose that $Q(X)$ has degree m.
   (b) Find a solution basis for the equation $P(D)Q(D)y = 0$. It will have $m + n$ constituent functions, and we may arrange things so that n of them satisfy $P(D)y = 0$.
   (c) Eliminate from the basis of step (b) those n functions that satisfy $P(D)y = 0$.
   (d) A particular solution exists that is a linear combination of the m functions that remain after step (c).

   You should be able to check that this gives the same substitutions in the case that g is a pure exponential polynomial (one of the form $f(x)e^{\lambda x}$) as use of Proposition 1.19. See if you like it by trying it out on the problem

$$y'' + y = \sin x + x\cos 2x.$$

4. Given an $m \times m$-matrix A the exponential e^A can be defined by the series

$$e^A = \sum_{k=0}^{\infty} \frac{1}{k!} A^k$$

   It can be shown that the series converges in the sense that each matrix element in the partial sum

$$\sum_{k=0}^{N} \frac{1}{k!} A^k$$

   converges as $N \to \infty$.

   (a) Now consider the operator $D : V_m^\lambda \to V_m^\lambda$ of differentiation in the m-dimensional space V_m^λ. Let a be a real number. What does the operator e^{aD} do to each function $f \in V_m^\lambda$?
   (b) Find a polynomial $y(x)$ that satisfies $y(x + 2) - y(x + 1) + y(x) = x^2$.

1.4.3 Projects

A. *Project on damped harmonic oscillations*
The equation

$$\frac{d^2y}{dt^2} + k\frac{dy}{dt} + \omega^2 y = 0, \quad (-\infty < t < \infty)$$

where $k \geq 0$ and $\omega > 0$, describes damped harmonic oscillations. It includes the theory of the RLC circuit in electrical engineering. Physically the constant k provides damping (resistance in electrical applications or friction in mechanical applications). We denote the independent variable by t, since, almost invariably in applications, it is time. Although we have posed the problem on the whole real line, the main interest lies in future time, $t > 0$.

A1. Show that every solution decays exponentially as $t \to \infty$; more precisely, there is a constant $\beta > 0$, depending only on k and ω, such that if $y(t)$ is a solution, then there exists $C > 0$, depending on the solution in question, such that

$$|y(t)| \leq Ce^{-\beta t}, \quad (t > 0).$$

In addition to decay, the oscillatory, or otherwise, behaviour of the solution can be important. In the next exercise we distinguish a number of cases, that are significant in engineering applications.

A2. (a) If $k = 0$ show that the general solution is periodic with period $2\pi/\omega$.
(b) Let $0 < k < 2\omega$. Show that the general solution is oscillatory, meaning that it has infinitely many zeros. Show that the difference between consecutive zeros is a constant and calculate it.

Oscillations in a mechanical system can be undesirable; think of a car's suspension on a bumpy road. To suppress them we need k to be sufficiently large.

(c) Let $k = 2\omega$. Show that a non-identically zero solution either changes sign exactly once, or else not at all. Calculate the solution that satisfies the Cauchy conditions

$$y(0) = a, \quad \frac{dy}{dt}(0) = b.$$

Indicate in a sketch of the (a, b)-plane those initial values for which the corresponding solution changes sign. Why might you say that the property, that the solution nowhere changes sign, is exceptional?
(d) Let $k > 2\omega$. Putting the zero solution aside show that the other solutions can be divided into two disjoint classes: those that never take the value 0

and those that take the value 0 exactly once, and that, unlike in the case of item (c), neither class is exceptional.

Note The cases (b), (c) and (d) are sometimes called *underdamped, critically damped* and *overdamped* respectively.

A3. Consider the equation for damped harmonic oscillations with a sinusoidal forcing term:

$$\frac{d^2y}{dt^2} + k\frac{dy}{dt} + \omega^2 y = F\cos\lambda t$$

where $k > 0$, $\omega > 0$, $F \neq 0$ and $\lambda \neq 0$.

(a) Show that there is a unique periodic solution $y = \phi(t)$ and calculate it.
(b) Let $\psi(t)$ be any other solution. Show that, as $t \to \infty$:

$$\psi(t) - \phi(t) \to 0$$

with exponential decay to 0 (see Exercise A1 for this notion).

This makes the solution $\phi(t)$ an object of practical interest. The solution $\psi(t)$ may be indistinguishable from $\phi(t)$ for moderately large t.

(c) For a given ω, determine λ (that is, adjust the input frequency) in order to maximise the amplitude of $\phi(t)$.
(d) For a given λ determine ω (that is, tune the oscillator) in order to maximise the amplitude of $\phi'(t)$.

B. *Project on Schrödinger's equation or a brief first course in quantum theory*
In physics the equation

$$\psi'' + (E - U(x))\psi = 0 \tag{1.23}$$

is called the stationary, one-dimensional Schrödinger equation. It describes a quantum-mechanical particle with energy E moving on the x-axis under a force field with potential U. The number E is a real constant and U a continuous, real-valued function. The use of the letter 'ψ' for the dependent variable is conventional in quantum theory.

If $U = 0$ and $E = k^2$ we have the problem for a free particle with momentum k

$$\psi'' + k^2\psi = 0. \tag{1.24}$$

The equation has a solution basis comprising e^{ikx} and e^{-ikx}. The first solution is called "the particle moving to the right with momentum k" and the second "the particle moving to the left with momentum k".

Now we suppose that U is not the zero-function but has compact support, that is, it is different from zero only in some bounded interval $K = [-L, L]$. If $U \leq 0$ it is called a potential well; if $U \geq 0$ a potential barrier. Also we assume that the energy E in (1.23) is positive and write $E = k^2$.

B1. Any solution ψ of (1.23) is a solution of (1.24) outside the interval K. To the left of K it can be written $Ae^{ikx} + Be^{-ikx}$ and to the right $Ce^{ikx} + De^{-ikx}$, for certain constants A, B, C and D. Prove that the vector (A, B) determines the vector (C, D) linearly, that is, there is a matrix M, such that

$$\begin{bmatrix} C \\ D \end{bmatrix} = M \begin{bmatrix} A \\ B \end{bmatrix}.$$

and that M has the form

$$M = \begin{bmatrix} \alpha & \beta \\ \bar{\beta} & \bar{\alpha} \end{bmatrix}$$

where

$$|\alpha|^2 - |\beta|^2 = 1.$$

B2. Show that the set of all matrices of this form make up a group (under multiplication) that may be described as the group of all 2×2 complex matrices that preserve the indefinite Hermitian form $|z_1|^2 - |z_2|^2$ (defined on the space $\mathbb{C}^2$) and have determinant 1. It is denoted by $SU(1, 1)$.
Hint Write the algebraic conditions that a matrix $\begin{bmatrix} \alpha & \beta \\ \gamma & \delta \end{bmatrix}$ preserves the form $|z_1|^2 - |z_2|^2$ (there should be three such) and then throw the condition $|\alpha|^2 - |\beta|^2 = 1$ into the mix.
Note The numbers '1,1' refer to the number of positive terms and negative terms in the indefinite form $|z_1|^2 - |z_2|^2$.

B3. Show that there is a unique solution ψ which is of the form $e^{ikx} + Ae^{-ikx}$ to the left of the interval K and of the form Be^{ikx} to the right of K. Express A and B in terms of α and β and show that $|A|^2 + |B|^2 = 1$.
Note This solution is interpreted as a particle with energy k^2 that comes in from the left, is reflected back with probability $|A|^2$ and is transmitted to the right ("penetrates the barrier") with probability $|B|^2$.

B4. Consider again an arbitrary solution ψ equal to $Ae^{ikx} + Be^{-ikx}$ to the left of K and $Ce^{ikx} + De^{-ikx}$ on the right. Show that the vector (A, D) determines the vector (B, C); that is, find a matrix S so that

$$\begin{bmatrix} B \\ C \end{bmatrix} = S \begin{bmatrix} A \\ D \end{bmatrix}$$

and express the elements of S in terms of α and β.

Note The matrix S is called the scattering operator. Physically (A, D) describes particles approaching K from left and right whilst (B, C) describes how they rebound from K.

B5. Show that S is a unitary matrix with determinant 1; that is if

$$S = \begin{bmatrix} a & b \\ c & d \end{bmatrix}$$

then $|a|^2 + |c|^2 = |b|^2 + |d|^2 = 1$, $a\bar{b} + c\bar{d} = 0$ and $ad - bc = 1$. The 2×2 unitary matrices form a group $U(2)$ which may be described as the group of operators that preserve the form $|z_1|^2 + |z_2|^2$. Those with determinant 1 form a subgroup denoted by $SU(2)$.

1.5 Boundary Value Problems

As the general solution of (1.8) contains n arbitrary constants it is plausible that n conditions (which we can call side conditions, as they stand in addition to the differential equation) imposed on the unknown function $y(x)$ might determine those constants and produce a unique solution. We have already seen one such example, that of initial conditions at a point x_0 in the interval I.

The general solution provided by Proposition 1.8 is actually an affine function of $c_1, \ldots, c_n$, namely

$$y(x) = c_1 u_1(x) + \cdots + c_n u_n(x) + v(x).$$

The initial values $y^{(k-1)}(x_0)$, $(k = 1, \ldots, n)$ are linear functionals on the space of functions n-times differentiable in the interval I. We could try replacing them by different, at present unspecified, linear functionals $U_k(y)$, $(k = 1, \ldots, n)$, and then use the conditions $U_k(y) = a_k$, $(k = 1, \ldots, n)$. This involves solving the equations

$$\sum_{j=1}^{n} c_j U_k(u_j) + U_k(v) = a_k, \quad (k = 1, \ldots, n).$$

It follows immediately that the necessary and sufficient condition that a unique solution exists for every choice of the continuous function $g(x)$, and every choice of the values $a_1, \ldots, a_n$, is that the determinant

$$\left| \left(U_k(u_j) \right)_{k=1}^{n} {}_{j=1}^{n} \right|$$

is not zero. Equivalently, it is that the restrictions to the span of the n functions $u_1, \ldots, u_n$, of the n linear functionals $U_1, \ldots, U_n$, are linearly independent (for which it is necessary, but not sufficient, that they are already linearly independent on the space of n-times differentiable functions). However, this assertion is usually expressed in the following way, which does not mention a solution basis:

Proposition 1.23 *Let the coefficient functions $p_0, \ldots, p_{n-1}$ be continuous in the open interval I. Then the non-homogeneous equation*

$$y^{(n)} + p_{n-1}(x)y^{(n-1)} + \cdots + p_1(x)y' + p_0(x)y = g(x), \quad (x \in I) \tag{1.25}$$

with the side conditions

$$U_1(y) = a_1, \quad \ldots, \quad U_n(y) = a_n,$$

has a unique solution for every choice of the continuous function g and every choice of the constants $a_1, \ldots, a_n$, if and only if the corresponding homogeneous problem

$$y^{(n)} + p_{n-1}(x)y^{(n-1)} + \cdots + p_1(x)y' + p_0(x)y = 0, \tag{1.26}$$

with the homogeneous side conditions

$$U_1(y) = 0, \quad \ldots \quad U_n(y) = 0,$$

has as its sole solution $y = 0$.

The calculations in the following example, illustrating Proposition 1.23, will resurface in Chap. 7, but with a different objective.

Example For which values of the constant λ does the problem

$$y'' + \lambda y = g(x), \quad y(0) = c_1, \ y(1) = c_2$$

have a unique solution for every choice of the continuous function g and the values c_1 and c_2?

Solution If $\lambda \neq 0$ we can use the basis $u_1(x) = e^{i\sqrt{\lambda}x}$, $u_2(x) = e^{-i\sqrt{\lambda}x}$. The condition that a unique solution to the inhomogeneous problem must always exist can be expressed by

$$\begin{vmatrix} 1 & 1 \\ e^{i\sqrt{\lambda}} & e^{-i\sqrt{\lambda}} \end{vmatrix} \neq 0$$

which is equivalent to

$$e^{2i\sqrt{\lambda}} \neq 1.$$

The solutions to the equation $e^{2i\sqrt{\lambda}} = 1$ are $\lambda = m^2\pi^2$ where m is an arbitrary integer, distinct from 0, since we excluded the value $\lambda = 0$ initially. To examine the possibility $\lambda = 0$ we need to use a basis appropriate to the case that the indicial equation has a double root, such as $u_1(x) = 1$, $u_2(x) = x$. In this case the solution $A + Bx$ only satisfies the homogeneous boundary conditions if $A = B = 0$. The answer is therefore: all λ except those of the form $m^2\pi^2$ where m is a non-zero integer.

More clarity about what is going on in Proposition 1.23 can be achieved by thinking in terms of vector spaces and linear algebra. This also furnishes an alternative proof of the proposition. In what follows we shall introduce vector spaces of functions in the interval I. If the differential equation has real valued coefficient functions we may use vector spaces over $\mathbb{R}$, comprising real valued functions. We may also use vector spaces over $\mathbb{C}$ of complex valued functions; and we must do so if the coefficient functions are complex valued. We shall not explicitly say whether the field is $\mathbb{R}$ (with real valued functions) or $\mathbb{C}$ (with complex valued functions), as the same conclusions hold in both cases.

Let Y_1 denote the space of all functions twice differentiable in the open interval I, and let Y_2 denote the space of functions continuous in the interval I. We can define the linear operator

$$L : Y_1 \to Y_2$$

by

$$(Ly)(x) = y^{(n)}(x) + p_{n-1}(x)y^{(n-1)}(x) + \cdots + p_1(x)y'(x) + p_0(x)y(x)$$

for each $y \in Y_1$.

The fundamental propositions 1.4 and 1.8 assert that L is surjective and its nullspace S is n-dimensional. Instead of introducing boundary conditions we let E be a subspace of Y_1 having codimension n in Y_1. For example E could be the subspace

$$E = \{y \in Y_1 : U_k(y) = 0,\ k = 1, \ldots, n\}$$

where $U_1, \ldots, U_n$ are n linearly independent linear functionals defined on Y_1 (defining n independent side conditions). In fact Y_1 can always be represented in this way for a suitable choice of the n linear functionals. The null-space of $L|_E$ (the restriction of L to E) is the intersection $E \cap S$, and can have any dimension from 0 up to n. The space $E \cap S$ comprises all solutions of $Ly = 0$ that satisfy the homogeneous side conditions $U_k(y) = 0$, $k = 1, \ldots, n$. But some basic linear algebra shows that *the dimension of this space is the same as the codimension in Y_2 of the range, $L(E)$, of $L|_E$.*

Given that the space of solutions of $Ly = 0$, that satisfy the n side conditions, has dimension r, (where $0 \le r \le n$), there must therefore exist r independent linear functionals $V_1, \ldots, V_r$, defined on the space Y_2 (not, of course, unique), such that the problem

$$Ly = g(x), \quad U_k(y) = 0, \quad (k = 1, \ldots, n)$$

has a solution for a given $g \in Y_2$ if and only if $V_k(g) = 0$ for $k = 1, \ldots, r$. It is an interesting and often important problem to determine suitable functionals V_k. Some examples of this are explored in the exercises in this section (see Exercises 10 and 11).

We give a brief summary of the proof by linear algebra of the claims of the last two paragraphs. It involves quotient spaces and maybe a bit more mathematical sophistication than we mostly require in this text. The reader should check the claims made, including that the mappings introduced are well defined. The mapping $L : Y_1 \to Y_2$ induces a mapping

$$\tilde{L} : \frac{Y_1}{E+S} \to \frac{Y_2}{L(E)}$$

given by[4]

$$\tilde{L}\big(u + (E+S)\big) = Lu + L(E).$$

Since L is surjective, $\tilde{L}$ is a linear bijection. To establish that the codimension of $L(E)$ in Y_2 is r, the same as the dimension of $E \cap S$, or the space of all solutions of $Ly = 0$ that lie in E, we want to show that the dimension of $Y_1/(E+S)$ equals r. We have what algebraists call a *short exact sequence* in which the arrows represent linear mappings

$$0 \to \frac{S}{S \cap E} \to \frac{Y_1}{E} \to \frac{Y_1}{E+S} \to 0.$$

The extreme mappings are 0. The second takes $y + (S \cap E)$ (where $y \in S$) to $y + E$ and is injective. The third takes $y + E$ to $y + (E+S)$ and is surjective. The reader may check that the range of each linear mapping is the null space of the next. All the spaces in the sequence are finite dimensional; of the three inner spaces the first has dimension $n - r$ and the second has dimension n. Applying the rank-nullity rule, that rank plus nullity of a linear mapping equals the dimension of the domain, we find that the dimension of $Y_1/(E+S)$ equals the dimension of Y_1/E minus the dimension of $S/(S \cap E)$, that is, it is r.

This manner of clothing the problem in the guise of linear algebra, in terms of spaces Y_1 and Y_2, and an operator L, is very useful. In particular, the definitions of Y_1 and Y_2 can be varied somewhat, without losing the essential structure described in

[4] We think of elements of a quotient as translates of the divisor subspace.

the preceding paragraphs. For example, the functions admitted may have boundary values. This is the content of the next section.

1.5.1 Boundary Conditions

In practical applications, while it is common to pose a differential equation in an open interval $]a, b[$, we may employ side conditions that involve the values of $y(x)$ and its derivatives up to order $n - 1$ at the endpoints a and b. The conditions are then called *boundary conditions*. For this to make sense the function $y(x)$ and its derivatives should extend to continuous functions on the closed interval $[a, b]$. This can be accomplished by simply supposing that the coefficient functions p_0, $\dots$, p_{n-1} and the inhomogeneous term g are continuous in the closed interval $[a, b]$. That $y(x)$ then has boundary values can be seen by extending the coefficient functions continuously outside the interval $[a, b]$, and applying the fundamental theorem, Proposition 1.4, to the larger interval. In practical applications, points outside $[a, b]$ are irrelevant, because, for example, the interval models a physical medium, such as a rod conducting heat, or a vibrating string.

In the parlance of the previous section, we are restricting the definitions of Y_1 and Y_2. For Y_1 we now use the space of functions with derivatives in $]a, b[$ up to order n, such that the function and its derivatives extend to continuous functions in the closed interval $[a, b]$. This is an important vector space of functional analysis, denoted by $C^n[a, b]$. For Y_2 we now use the space of functions continuous in the closed interval $[a, b]$. This space is commonly denoted by $C[a, b]$ and is perhaps the most important function space of functional analysis.

As two examples of boundary conditions we can name the initial conditions at a,

$$y(a) = c_1, \quad \dots, \quad y^{(n-1)}(a) = c_n,$$

and the initial conditions at b,

$$y(b) = c_1, \quad \dots, \quad y^{(n-1)}(b) = c_n.$$

By the fundamental theorem, both these sets of boundary conditions will determine a unique solution in $]a, b[$.

More generally, boundary conditions will involve a family of n linear combinations of the $2n$ quantities $y^{(k-1)}(a)$ and $y^{(k-1)}(b)$, $(k = 1, \dots, n)$. They should be linearly independent linear functionals of their $2n$ variables, for otherwise we do not genuinely have n side conditions and there is no hope that they determine a unique solution. A boundary condition is called homogeneous when it is satisfied by the zero-function.

Let us look at commonly occurring types of boundary conditions for the important case of a second order equation

$$y'' + p(x)y' + q(x)y = g(x), \quad (a < x < b).$$

We assume that the functions p, q and g are continuous in the closed interval $[a, b]$. In addition to initial conditions, posed either entirely at a or entirely at b, the following two types are probably the most important because of their practical applications:

Separated Boundary Conditions These are of the form

$$\alpha_1 y(a) + \alpha_2 y'(a) = c_1$$
$$\beta_1 y(b) + \beta_2 y'(b) = c_2$$

For linear independence we require that neither (α_1, α_2) nor (β_1, β_2) is the zero-vector $(0, 0)$. The conditions are homogeneous when $c_1 = c_2 = 0$.

Periodic Boundary Conditions These are of the form

$$y(a) = y(b)$$
$$y'(a) = y'(b)$$

and are generally imposed as homogeneous conditions (that is, they are satisfied by the zero function), as shown here.

1.5.2 Green's Function

We are concerned with the problem treated in Proposition 1.23, on an interval $I :=]a, b[$, with boundary conditions at a and b. We therefore assume first of all that the coefficient functions $p_0, \ldots, p_{n-1}$ are continuous in the closed interval $[a, b]$ and consider the homogeneous problem

$$\begin{aligned} y^{(n)} + p_{n-1}(x)y^{(n-1)} + \cdots + p_1(x)y' + p_0(x)y = 0, \quad (a < x < b) \\ U_1(y) = 0, \quad \ldots \quad U_n(y) = 0. \end{aligned} \tag{1.27}$$

We are assuming general boundary conditions in which $U_k(y)$ is a linear combination of the $2n$ quantities

$$y(a), y'(a), \ldots, y^{(n-1)}(a), y(b), y'(b), \ldots, y^{(n-1)}(b).$$

Suppose now that the homogeneous problem has as sole solution $y = 0$. Then the non-homogeneous problem has a unique solution. We are going find a representation of the solution in the form of an integral transform. This is analogous to expressing the solution vector of a system of linear equations using the inverse of the coefficient matrix.

Although the problem (1.27) has no non-zero solution, we can find a non-zero function that "almost satisfies it," by allowing it to have discontinuities.

Definition A Green's function[5] for the problem (1.27) is a function $G(x, \xi)$ of two variables, $a \leq x \leq b$, $a < \xi < b$, with the following properties:

1. As a function of x, $G(x, \xi)$ satisfies the differential equation in (1.27) for $a < x < \xi$ and for $\xi < x < b$.
2. As a function of x, $G(x, \xi)$ satisfies all the boundary conditions in (1.27).
3. The partial derivatives $D_1^j G(x, \xi)$, for $j = 0, \ldots, n-2$, exist and are continuous functions of x for each ξ.
   Note The operator of partial differentiation, written here as D_1, is intended to differentiate the following function with respect to its first argument. This notation is preferred to the alternative D_x because 'x' in $D_1 G(x, \xi)$, or its higher derivatives, will often be replaced by another expression, as in the next item.
4. The partial derivative $D_1^{n-1} G(x, \xi)$ is a continuous function of x for $a < x < \xi$ and for $\xi < x < b$. At $x = \xi$ it has a jump discontinuity, such that

$$D_1^{n-1} G(\xi+, \xi) - D_1^{n-1} G(\xi-, \xi) = 1.$$

It is a fact that under the stated conditions a Green's function exists and is unique.

Proposition 1.24 *Assume that the homogeneous problem* (1.27) *has as sole solution $y = 0$. Then conditions 1 to 4 determine a Green's function uniquely.*

Proof Let $u_1, \ldots, u_n$ be a solution basis for the differential equation in (1.27). Then the restrictions of $u_1, \ldots, u_n$ to any open subinterval of $]a, b[$ form a solution basis for the equation on that interval. Hence there exist $A_1, \ldots, A_n$ and $B_1, \ldots, B_n$, depending on ξ, such that the Green's function, if it exists, will satisfy

$$G(x, \xi) = \sum_{j=1}^{n} A_j u_j(x), \quad (a < x < \xi)$$

$$G(x, \xi) = \sum_{j=1}^{n} B_j u_j(x), \quad (\xi < x < b).$$

[5] The ungrammatical formation is customary. One also says "the Green's function". An alternative is to say "the Green function", compare "the Euler characteristic". However this sounds disconcerting, perhaps because "green" is a commonplace English adjective.

Conditions 3 and 4 will be satisfied if

$$\sum_{j=1}^{n}(A_j - B_j)u_j^{(k)}(\xi) = \begin{cases} 0, & (k = 0, \ldots, n-2) \\ -1, & (k = n-1) \end{cases}$$

Because the Wronskian of the solutions $u_1, \ldots, u_n$ is non-zero, these equations determine $A_k - B_k$, for $k = 1, \ldots, n$, uniquely, and they are continuous functions of ξ.

It remains to consider the boundary conditions. Write

$$U_k(y) = U_k^L(y) + U_k^R(y)$$

where $U_k^L(y)$ comprises all the boundary terms that involve the left-hand endpoint a, whilst $U_k^R(y)$ comprises all terms involving the right-hand endpoint b. Then the boundary conditions require

$$\sum_{j=1}^{n} A_j U_k^L(u_j) + \sum_{j=1}^{n} B_j U_k^R(u_j) = 0, \quad (k = 1, \ldots, n)$$

or, equivalently,

$$\sum_{j=1}^{n} A_j (U_k^L(u_j) + U_k^R(u_j)) + \sum_{j=1}^{n}(B_j - A_j)U_k^R(u_j) = 0, \quad (k = 1, \ldots, n)$$

that is,

$$\sum_{j=1}^{n} A_j U_k(u_j) = \sum_{j=1}^{n}(A_j - B_j)U_k^R(u_j) = 0, \quad (k = 1, \ldots, n)$$

The assumptions about the homogeneous problem imply that the determinant

$$\left|\left(U_k(u_j)\right)_{j=1}^{n}{}_{k=1}^{n}\right|$$

is not zero. Therefore this system of equations determines A_j (for $j = 1, \ldots, n$) in terms of the already known quantities $A_j - B_j$, ($j = 1, \ldots, n$). Finally, therefore, both A_j and B_j are determined as functions, $A_j(\xi)$ and $B_j(\xi)$ of ξ. □

The proof of the previous proposition reveals that the Green's function $G(x, \xi)$ has more regularity than is assumed in its definition. The solutions $u_1, \ldots, u_n$ have continuous derivatives of order up to n, and these extend to continuous functions on the closed interval $[a, b]$. It follows that $A_j(\xi)$ and $B_j(\xi)$ extend to continuous functions on the closed interval $a \leq \xi \leq b$. Therefore the following stronger versions of properties 3 and 4 are valid:

3′. The partial derivatives $D_1^j G(x, \xi)$, for $j = 0, \ldots, n-2$, extend to continuous functions of (x, ξ) in the closed rectangle $a \leq x \leq b, a \leq \xi \leq b$.
4′. The partial derivative $D_1^{n-1} G(x, \xi)$ restricted to the open triangle $a < x < \xi$, $a < \xi < b$ extends to a continuous function of (x, ξ) in the closed triangle $a \leq x \leq \xi, a \leq \xi \leq b$. Similarly the partial derivative $D_1^{n-1} G(x, \xi)$ in the open triangle $\xi < x < b, a < \xi < b$ extends to a continuous function of (x, ξ) in the closed triangle $\xi \leq x \leq b, a \leq \xi \leq b$.

The additional regularity described here is needed to justify the calculations of the next proposition. Note that by property 4′ the left and right-hand limits in property 4 are no longer constrained to fixed ξ. The function $D_1^{n-1} G(x, \xi)$ has a genuine jump discontinuity as (x, ξ) crosses the line $x = \xi$.

The proposition shows how the Green's function provides a solution formula for the problem in which the differential equation is inhomogeneous but the boundary conditions remain homogeneous.

Proposition 1.25 *Under the conditions of Proposition 1.24, and given the function g continuous in the interval $[a, b]$, the unique solution of the boundary value problem*

$$\begin{gathered} y^{(n)} + p_{n-1}(x) y^{(n-1)} + \cdots + p_1(x) y' + p_0(x) y = g(x), \quad (a < x < b) \\ U_1(y) = 0, \quad \ldots, \quad U_n(y) = 0. \end{gathered} \tag{1.28}$$

is given by

$$y(x) = \int_a^b G(x, \xi) g(\xi)\, d\xi, \quad (a < x < b).$$

Proof Let $y(x)$ be defined by the stated formula. The derivatives of $y(x)$ up to order $n-2$ may be computed across the integral sign, thus,

$$y^{(k)}(x) = \int_a^b D_1^k G(x, \xi) g(\xi)\, d\xi, \quad (a < x < b,\ k = 0, 1, \ldots, n-2). \tag{1.29}$$

This is standard, because the derivatives $D_1^k G(x, \xi)$ in question are continuous functions of (x, ξ) (property 3′). The $(n-1)$st and nth derivatives require care. We write

$$y^{(n-2)}(x) = \int_a^x D_1^{n-2} G(x, \xi) g(\xi)\, d\xi + \int_x^b D_1^{n-2} G(x, \xi) g(\xi)\, d\xi$$

and compute the derivative, using the fundamental theorem of calculus and differentiation across the integral sign (actually requiring properties 3′ and 4′), as follows:

$$\begin{aligned} y^{(n-1)}(x) = \lim_{\substack{\xi \to x \\ \xi < x}} D_1^{n-2} G(x, \xi) g(x) - \lim_{\substack{\xi \to x \\ \xi > x}} D_1^{n-2} G(x, \xi) g(x) \\ + \int_a^x D_1^{n-1} G(x, \xi) g(\xi)\, d\xi + \int_x^b D_1^{n-1} G(x, \xi) g(\xi)\, d\xi \\ = \int_a^b D_1^{n-1} G(x, \xi) g(\xi)\, d\xi \end{aligned} \tag{1.30}$$

the limits cancelling out since the derivative $D_1^{n-2} G(x, \xi)$ is continuous across the line $x = \xi$. We use the same idea for the next derivative. This time we get

$$\begin{aligned} y^{(n)}(x) = \lim_{\substack{\xi \to x \\ x < \xi}} D_1^{n-1} G(x, \xi) g(x) - \lim_{\substack{\xi \to x \\ x > \xi}} D_1^{n-1} G(x, \xi) g(x) \\ + \int_a^b D_1^n G(x, \xi) g(\xi)\, d\xi \\ = g(x) + \int_a^b D_1^n G(x, \xi) g(\xi)\, d\xi \end{aligned} \tag{1.31}$$

by property 4′ of the Green's function. Using next property 1 we find

$$y^{(n)} + p_{n-1}(x) y^{(n-1)} + \cdots + p_1(x) y' + p_0(x) y = g(x), \quad (a < x < b)$$

and by property 2 we have that $y(x)$ satisfies the homogeneous boundary conditions $U_k(y) = 0$. □

Practicalities

We present some practical hints for the calculation of the Green's function for what is probably the most important case, that of the second order equation, with separated boundary conditions:

$$\begin{aligned} y'' + p(x) y' + q(x) y = g(x), \quad (a < x < b) \\ \alpha_1 y(a) + \alpha_2 y'(a) = c_1, \quad \beta_1 y(b) + \beta_2 y'(b) = c_2 \end{aligned} \tag{1.32}$$

As usual we assume that $(\alpha_1, \alpha_2) \neq (0, 0)$ and $(\beta_1, \beta_2) \neq (0, 0)$.

As a general rule one can carry out the procedure used in the proof of Proposition 1.24. However, for separated boundary conditions it simplifies matters to choose a basis of solutions u_1, u_2, such that u_1 satisfies the homogeneous boundary condition at $x = a$ (that is, with $c_1 = 0$), whilst u_2 satisfies the homogeneous boundary condition at $x = b$. If the homogeneous problem (the one with $g = 0$, $c_1 = c_2 = 0$) admits only the zero solution (which is the condition that the Green's function should exist), functions u_1 and u_2 chosen by this prescription (provided they are not identically zero) are linearly independent (for otherwise each would be a solution of the homogeneous problem), and hence form a basis. Then the Green's function will have the form

$$G(x, \xi) = \begin{cases} Au_1(x), & (a < x < \xi < b) \\ Bu_2(x), & (a < \xi < x < b) \end{cases}$$

This already satisfies the boundary conditions. We only have to find A and B (as functions of ξ) in order to satisfy the continuity conditions at $x = \xi$:

$$\begin{aligned} Au_1(\xi) - Bu_2(\xi) &= 0, \\ -Au_1'(\xi) + Bu_2'(\xi) &= 1 \end{aligned}$$

A further trick is possible when the equation has constant coefficients, and the boundary conditions are "the same" at both ends. By the latter we mean that the problem has the form

$$\begin{gathered} y'' + py' + qy = g(x), \quad (a < x < b) \\ \alpha_1 y(a) + \alpha_2 y'(a) = c_1, \quad \alpha_1 y(b) + \alpha_2 y'(b) = c_2 \end{gathered} \tag{1.33}$$

In this case, having chosen u_1 to satisfy the homogeneous boundary condition at $x = a$, we can choose for u_2 the function

$$u_2(x) := u_1(x + a - b).$$

The reason for this is that a translate of a solution of an equation with constant coefficients is again a solution. Sometimes this leads to some simplification in the calculation of the Green's function.

1.5.3 Exercises

1. Show that the problem

$$\begin{gathered} y''' = g(x), \quad (0 < x < 1) \\ y(0) = c_1, \ y''(0) = c_2, \ y'(1) = c_3 \end{gathered}$$

has a unique solution for all choices of the continuous function g and the values c_1, c_2 and c_3.

2. Find the condition that the *complex constant* λ must satisfy in order that the inhomogeneous problem

$$y'' + \lambda y = g(x), \quad (0 < x < 1)$$
$$y'(0) = c_1, \quad y'(1) = c_2$$

has a unique solution for all choices of the continuous function g and the values c_1 and c_2.

3. Find the condition that the constants p and q must satisfy in order that the inhomogeneous problem

$$y'' + py' + qy = g(x), \quad (a < x < b)$$
$$y(a) = c_1, \quad y(b) = c_2$$

has a unique solution for all choices of the continuous function g and the values c_1 and c_2.
Hint Express the condition first in terms of the roots λ_1, λ_2 of the indicial equation, assumed distinct. Then rewrite it in terms of p and q. Finally treat the case $\lambda_1 = \lambda_2$.

4. For the problem (1.32), and using the basis explained in the text, show that

$$G(x,\xi) = \begin{cases} W(\xi)^{-1}u_1(x)u_2(\xi), & (a < x < \xi < b) \\ W(\xi)^{-1}u_2(x)u_1(\xi), & (a < \xi < x < b) \end{cases}$$

where W is the Wronskian of u_1 and u_2. Show that the solution of (1.32) in the case that $c_1 = c_2 = 0$ is:

$$y(x) = u_2(x)\int_a^x W^{-1}u_1 g + u_1(x)\int_x^b W^{-1}u_2 g$$

How should this be modified in the case that c_1 and c_2 are not both 0?

5. Compute the Green's function $G(x, \xi)$ for the following boundary value problems, and hence express the solution for each problem in the form

$$y(x) = \int_a^b G(x,\xi)h(\xi)\,d\xi.$$

(a) $y'' = h(x), \quad y(a) = 0,\ y(b) = 0.$
(b) $y'' + y = h(x), \quad y(0) = 0,\ y(\frac{\pi}{2}) = 0.$
(c) $y'' + y = h(x), \quad y(0) = y(\frac{\pi}{2}),\ y'(0) = y'(\frac{\pi}{2}).$

6. Find the condition that L and k must satisfy that is necessary and sufficient for the problem

$$y'' + y = h(x), \quad (0 < x < L),$$
$$y(0) = 0, \quad y(L) - ky'(L) = 0$$

to have a Green's function, and compute the latter, when it exists.
Hint Let $u_1(x)$ be a solution of the homogeneous problem that satisfies $u_1(0) = 0$, and let $u_2(x) = v(x - L)$, where $v(x)$ is a solution of the homogeneous equation that satisfies $v(0) - kv'(0) = 0$.
7. Find the Green's function $G(x, \xi)$ for the problem

$$y^{(n)} = h(x), \quad (a < x < b),$$
$$y^{(k)}(a) = 0, \quad (k = 0, \ldots, n - 1)$$

Write a formula for the solution given that $h(x)$ is continuous in the interval $[a, b]$.
8. Consider the equation

$$y'' + p(x)y' + q(x)y = 0, \quad (-L < x < L),$$

on the assumption that $p(x)$ is an odd function and $q(x)$ an even function.

(a) Show that there is a basis for the solution space, consisting of functions $u_1(x)$, $u_2(x)$, such that u_1 is an odd function whilst u_2 is an even function.
(b) Consider the problem together with the separated boundary conditions

$$-\alpha y'(L) + \beta y(L) = 0, \quad \alpha y'(-L) + \beta y(-L) = 0,$$

where $(\alpha, \beta) \neq (0, 0)$. Suppose that $\phi(x)$ is a non-identically zero solution that satisfies these boundary conditions. Show that either ϕ is an odd function or it is an even function.

Hint In both items exploit the fact that if $y(x)$ satisfies the equation, so does $y(-x)$.
9. Consider the boundary value problem with non-constant leading coefficient

$$p_n(x)y^{(n)} + p_{n-1}(x)y^{(n-1)} + \cdots + p_0(x)y = h(x), \quad (a < x < b)$$
$$U_1(y) = 0, \quad \ldots, \quad U_n(y) = 0.$$

We assume that the coefficient functions $p_0, \ldots, p_n$ and the inhomogeneous term h are continuous in the closed interval $[a, b]$, and in addition that $p_n(x) \neq 0$ for $a \leq x \leq b$. For this problem we prescribe the Green's function $G(x, \xi)$ exactly as for the problem with $p_n = 1$, except that property 4 is modified by requiring that

$$D_1^{n-1}G(\xi+,\xi) - D_1^{n-1}G(\xi-,\xi) = 1/p_n(\xi).$$

Verify that Propositions 1.24 and 1.25 hold in this case with the obvious modifications to the premises.

10. We consider the boundary value problem

$$y'' + p(x)y' + q(x)y = g(x), \quad (a < x < b)$$

$$y(a) = c_1, \quad y(b) = c_2.$$

on the assumption that the homogeneous problem has a non-trivial (that is, not identically zero) solution $\phi(x)$.

(a) Let $\psi(x)$ be the solution of $y''+p(x)y'+q(x)y = 0$ that satisfies the initial conditions $\psi(a) = 1$, $\psi'(a) = 0$. Show that the functions ψ and ϕ form a solution basis for the equation $y'' + p(x)y' + q(x)y = 0$ in the interval $]a, b[$.

(b) Show that the boundary value problem has a solution if and only if c_1, c_2 and $g(x)$ satisfy a certain homogeneous linear constraint, which can be expressed in the form

$$\psi(b)\Big(c_1 + \int_a^b W^{-1}\phi\, g\Big) - c_2 = 0$$

where $W(x) := W(\phi, \psi)(x)$ is the Wronskian. Explain why the solution is not unique.

Hint One can refer to 1.2 Exercise 17, taking $x_0 = a$.

(c) Show that the function $w := W^{-1}\phi$, appearing in the integral in the previous item, satisfies the homogeneous boundary value problem

$$y'' - (p(x)y)' + q(x)y = 0, \quad (a < x < b)$$

$$y(a) = 0, \quad y(b) = 0.$$

11. Extend the results of the previous exercise to the more general boundary value problem

$$y'' + p(x)y' + q(x)y = g(x), \quad (a < x < b)$$

$$\alpha_1 y'(a) + \alpha_2 y(a) = c_1, \quad \beta_1 y'(b) + \beta_2 y(b) = c_2$$

assuming that neither (α_1, α_2) nor (β_1, β_2) is the zero vector $(0, 0)$, and that the homogeneous problem has a non-identically zero solution $\phi(x)$.

Hint It can help to use a solution basis ϕ, ψ such that ϕ is the stated solution of the homogeneous problem and ψ satisfies $\alpha_1\psi'(a) + \alpha_2\psi(a) = 1$. A conclusion like that of item (c) from the previous exercise will hold with the same differential equation, but care is needed with the boundary conditions.

Chapter 2
Separation of Variables

As you are now so once were we.

In this chapter we shall study mainly first order, usually non-linear, equations of the form $y' = f(x, y)$, with a view to obtaining solutions by quadratures (or integrals). In general such a solution method is not possible, but in certain cases, that are extremely important in practice, techniques going under the name of *separation of variables* can be applied. Later in the chapter we shall study second order equations that can be solved by reducing them to a family of first order equations, that, in turn, can be solved using separation of variables.

The function f is usually defined in an open connected subset D of the coordinate plane $\mathbb{R}^2$. By a solution of $y' = f(x, y)$ we will mean a function $\phi : I \to \mathbb{R}$, where I is an open interval, such that the graph of ϕ lies in D and $\phi'(x) = f(x, \phi(x))$ for all $x \in I$. The graph of a solution is called a *solution curve*. The *Cauchy problem*, with initial condition (or Cauchy condition) $y(x_0) = y_0$, asks for a solution curve that contains the point (x_0, y_0). In full generality, no method is known to obtain the solution by integrals.

In the first place we shall concern ourselves with a class of problems, the *separable equations*, that can be solved by means of two fundamental "inverse" operations that belong to analysis: that of obtaining an integral (more precisely an antiderivative) and that of obtaining an inverse function.

2.1 Separable Equations

The equation $y' = f(x, y)$ is said to be *separable* if

$$f(x, y) = h(x)g(y)$$

R. Magnus, *Essential Ordinary Differential Equations*, Springer Undergraduate Mathematics Series, https://doi.org/10.1007/978-3-031-11531-8_2

for certain functions h and g of one variable. The case when f does not depend on the independent variable x is of great practical importance. It is called the autonomous case. Although not logically necessary we will treat this case first, as the main features of separable equations appear most clearly in it.

2.1.1 The Autonomous Case

The equation $y' = f(x, y)$ is autonomous when f does not depend on the independent variable x. This means we have the problem

$$y' = g(y),$$

plainly a special type of separable equation. We assume throughout that $g(y)$ is *continuously differentiable* (or C^1—which is the same thing) in an open interval $]b_1, b_2[$, and that it has only isolated zeros. The domain of the equation is therefore a horizontal strip $D = \{(x, y) : b_1 < y < b_2\}$ in the plane $\mathbb{R}^2$.

Consider the Cauchy problem

$$y' = g(y), \quad y(x_0) = y_0.$$

To make sense, this requires that $b_1 < y_0 < b_2$. If y_0 is a zero of $g(y)$ the constant function $y(x) = y_0$ is a solution, and furthermore is the only solution of the Cauchy problem (more precisely it is the only solution on a given interval—a somewhat trivial distinction). This conclusion follows from the existence theory of Chap. 4 (looking ahead to Proposition 4.1), but we will also obtain it shortly by direct calculation. Uniqueness may fail if $g(y)$ is merely continuous, which is one reason why we assume that $g(y)$ is C^1.

Suppose next that $g(y_0) \neq 0$. To fix our ideas we can suppose that $g(y_0) > 0$. Again, by the existence theory to be presented in Chap. 4, a unique solution exists on an interval containing x_0. However, we shall derive this conclusion here and now, by the process of "solving the equation". This means that we assume a solution exists, obtain a formula for it and check that the formula gives a solution. This procedure, if successful, also implies that the solution is unique.

Let the solution be $y = \phi(x)$. Then, since $g(y_0) > 0$ and g is continuous, it follows that $\phi(x)$ is strictly increasing in a sufficiently small neighbourhood of x_0. There exists $h > 0$, such that the function $g(\phi(x))$ is positive in the interval $]x_0 - h, x_0 + h[$, and we have

$$\frac{\phi'(x)}{g(\phi(x))} = 1, \quad (x_0 - h < x < x_0 + h). \tag{2.1}$$

Let $G(y)$ be an antiderivative of $1/g(y)$ valid for a neighbourhood of y_0. Then

$$\frac{d}{dx}G(\phi(x)) = \frac{\phi'(x)}{g(\phi(x))} = 1, \quad (x_0 - h < x < x_0 + h)$$

so that

$$G(\phi(x)) = x + C, \quad (x_0 - h < x < x_0 + h) \tag{2.2}$$

for a certain constant C. By the Cauchy condition $C = G(y_0) - x_0$, so the solution satisfies

$$G(\phi(x)) = x - x_0 + G(y_0), \tag{2.3}$$

which we can also express in the form

$$\int_{y_0}^{\phi(x)} \frac{dy}{g(y)} = x - x_0. \tag{2.4}$$

To obtain the solution $y = \phi(x)$ we have to compute the inverse function of $y \mapsto G(y)$ in a neighbourhood of y_0. The same reasoning works if $g(y_0) < 0$.

The argument of the last paragraph produces a formula for the solution on the assumption that it exists. We can work backwards from this to show that the solution exists. It is simple to show by differentiation that the unique function $\phi(x)$ that satisfies (2.3) solves the Cauchy problem. Since the solution has to satisfy (2.3), uniqueness is obtained in the case that y_0 is not a zero of $g(y)$.

Now let's consider the extension of the solution. The purpose is to obtain a complete global picture of the solution curves.

First we suppose that y_0 lies between two consecutive zeros of $g(y)$. Suppose that $g(y_1) = g(y_2) = 0$ but $g(y) \neq 0$ for $y_1 < y < y_2$, and let $y_1 < y_0 < y_2$. To fix things, we suppose that $g(y) > 0$ for all $y \in \,]y_1, y_2[$.

As long its graph remains in the strip $y_1 < y < y_2$, the solution $y = \phi(x)$ is increasing and satisfies (2.4). Moreover the constants y_1 and y_2 are also solutions. We can ask whether the solution $\phi(x)$ can reach the value y_2 at some point $x = x_1$ to the right of x_0. If it can, then by (2.4) we have

$$\int_{y_0}^{y_2} \frac{dy}{g(y)} = x_1 - x_0.$$

The integral is improper at y_2 and in fact must diverge. For as $g(y)$ is continuously differentiable there exists a continuous function $h(y)$, positive for $y_0 < y < y_2$, such that

$$g(y) = (y_2 - y)h(y).$$

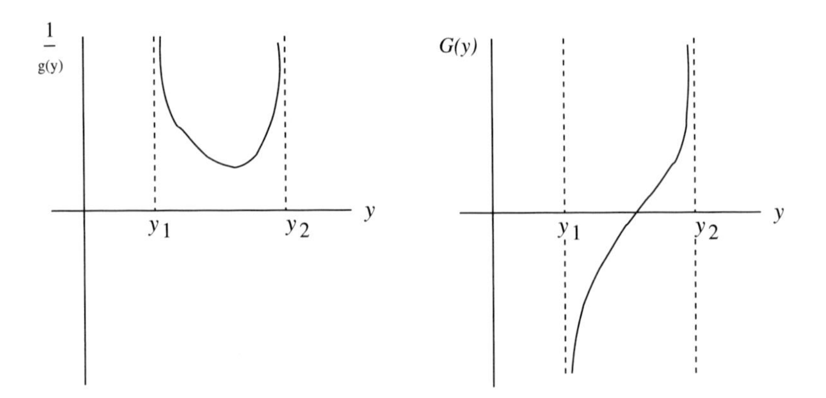

Fig. 2.1 Antiderivative of $1/g(y)$ between two zeros of $g(y)$

Then

$$\int_{y_0}^{y_2} \frac{dy}{g(y)} = \int_{y_0}^{y_2} \frac{dy}{(y_2 - y)h(y)} > \int_{y_0}^{y_2} \frac{dy}{M(y_2 - y)} = \infty,$$

where M is an upper bound for $h(y)$ on $[y_0, y_2]$. The argument shows that a finite point x_1 cannot exist. There is therefore no limit to how far the solution can be extended to the right and a similar argument applies to extending it to the left. See Fig. 2.1 for an illustration of the divergence of the integral.

To see this from a different point of view we can suppose that the solution has been extended to the right for $x < c$. Since $\phi(x)$ is increasing and bounded above by y_2 the limit $\eta := \lim_{x \to c-} \phi(x)$ exists and by the differential equation

$$\lim_{x \to c-} \phi'(x) = \lim_{x \to c-} g(\phi(x)) = g(\eta).$$

We can now solve the equation with Cauchy condition $y(c) = \eta$ and we obtain a solution, that glued to $y = \phi(x)$ extends the latter beyond $x = c$.

The conclusion here is that the solution exists on the interval $]-\infty, \infty[$ and its graph lies in the strip $y_1 < y < y_2$. We next consider what happens as x tends to ∞ or $-\infty$. In fact we show that $\lim_{x \to \infty} \phi(x) = y_2$. Since the solution is increasing and bounded above by y_2 it follows that the limit $\eta := \lim_{x \to \infty} \phi(x)$ exists and $\lim_{x \to \infty} \phi'(x) = 0$. By the differential equation we also have $\lim_{x \to \infty} \phi'(x) = g(\eta)$. We conclude that $g(\eta) = 0$ and since $y_1 < \eta \leq y_2$ we must have $\eta = y_2$, that is, $\lim_{x \to \infty} \phi(x) = y_2$. A similar argument shows that $\lim_{x \to -\infty} \phi(x) = y_1$.

In addition, the formulas (2.3) and (2.4) hold for $-\infty < x < \infty$. We should choose $G(y)$ to be an antiderivative of $1/g(y)$ valid for the interval $]y_1, y_2[$.

Similar reasoning disposes of the case when $g(y) < 0$ for $y_1 < y < y_2$. Now the solution is decreasing, tends to y_2 as $x \to -\infty$, to y_1 as $x \to \infty$, and is again given by (2.3) or (2.4).

Next we consider an initial condition that lies above the highest zero of $g(y)$. The simplest case is when $b_2 = \infty$. To fix the ideas let's suppose that $g(y_2) = 0$, that $g(y) > 0$ for $y_2 < y < \infty$ and that $y_2 < y_0$.

The solution is increasing. The reasoning of the previous paragraphs shows that the solution can be extended indefinitely to the left and approaches y_2 as $x \to -\infty$. Suppose the solution can be extended to the right for $x < c$. Then either the limit $\lim_{x\to c-} \phi(x)$ is ∞, so we have a vertical asymptote at $x = c$, or else the limit $\lim_{x\to c-} \phi(x)$ is finite. But in the latter case the solution can be extended further by the same argument as in the preceding paragraphs. So either the solution can be extended to all $x < \infty$ or else there is a vertical asymptote. Similar reasoning applies to the strip below the lowest zero of $g(y)$ if $b_1 = -\infty$. In the following exercise a simple criterion is given for the presence, or otherwise, of vertical asymptotes.

Exercise Let $g(y)$ be positive in the interval $]y_2, \infty[$ and let $y_2 < y_0$. Show that the solution of

$$y' = g(y), \quad y(0) = y_0$$

has a vertical asymptote if and only if the integral

$$\int_{y_0}^{\infty} \frac{dy}{g(y)}$$

is convergent. Similar conclusions, which the reader should frame, hold if $g(y)$ is negative, or the interval is of the form $]-\infty, y_1[$.

As a final touch we note that any translate of a solution is again a solution (prove this!). If $y = \phi(x)$ is a solution, so is $y = \phi(x - c)$. So it really suffices to find one solution curve in each interval stretching between consecutive zeros of $g(y)$. All others in the same strip are translates of the first one. The conclusions here are illustrated in Fig. 2.2 for an important example called the logistic equation.

What about the promised proof that the constant $y = y_0$ is the only solution of the Cauchy problem if $g(y_0) = 0$? If there exists another solution $y = \phi(x)$, not a constant but still satisfying $\phi(x_0) = y_0$, then we can find an open interval $]x_1, x_2[$ in which $\phi(x) \neq y_0$, and is such that either x_1 is finite and $\phi(x_1) = y_0$, or x_2 is finite and $\phi(x_2) = y_0$. But we have seen that a non-constant solution curve cannot reach a zero of $g(y)$ at any finite value of x. Note that the argument depends heavily on the differentiability of $g(y)$.

It is usual to refer to the equation

$$G(y) = x + C \tag{2.5}$$

(compare (2.2)) as the general solution. But care must be taken. Firstly, the antiderivative $G(y)$ of $1/g(y)$ will depend on which interval between consecutive zeros of $g(y)$ we are interested in. Secondly, the constant solutions $y = y_0$ occurring when y_0 is a zero of $g(y)$ are not of this form in general, that is, they cannot be obtained by assigning a value to C. Thirdly, the equation only determines y implicitly as a function of x. It always remains to find the inverse function of $G(y)$. More details of the arguments used here will be given in the next section, when we study the non-autonomous case.

Traditionally the derivation of (2.5) is carried out by the formal separation of x and y as follows. Write the equation as

$$\frac{dy}{dx} = g(y)$$

and "separate the variables"

$$\frac{dy}{g(y)} = dx$$

Hence

$$\int \frac{dy}{g(y)} = \int dx + C$$

and writing antiderivatives

$$G(y) = x + C.$$

Example The logistic equation is the problem

$$y' = y\left(1 - \frac{y}{L}\right) \tag{2.6}$$

where L is a positive constant. It is a popular model for the growth of a biological population that grows exponentially when the population is low but whose growth rate is curtailed by overcrowding. Similar equations are used to model pandemics, and have been put recently to good use.[1]

Solution of the Logistic Equation We have $g(y) = y\left(1 - \frac{y}{L}\right)$ with two zeros, $y = 0$ and $y = L$. These are therefore the only constant solutions. For $y < 0$ we have $g(y) < 0$, for $0 < y < L$ we have $g(y) > 0$ and for $L < y$ we have $g(y) < 0$. The integrals $\int^{\infty} 1/g$ and $\int^{-\infty} 1/g$ are convergent, revealing vertical asymptotes. We can immediately sketch the solution curves (Fig. 2.2).

Now for the calculations. The general solution is

$$\int \frac{dy}{y\left(1 - \frac{y}{L}\right)} = x + C$$

[1] This sentence was added in April 2021.

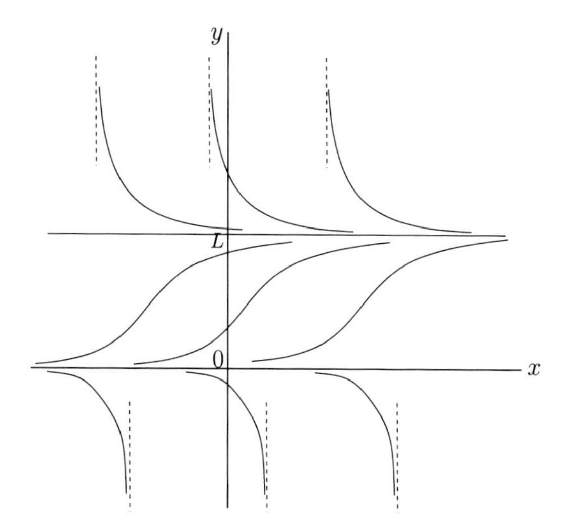

Fig. 2.2 Solution curves for the logistic equation $y' = y(1 - \frac{y}{L})$

or

$$\int \left(\frac{1}{L-y} + \frac{1}{y} \right) dy = x + C$$

How we write the antiderivative depends on the interval. For $y < 0$ we write

$$-\ln(L-y) + \ln(-y) = x + C$$

We then find

$$y = \frac{LAe^x}{Ae^x - 1}$$

where $A = e^C > 0$. This describes a decreasing solution in the region $y < 0$ defined in the interval $]-\infty, -\ln A[$, with a vertical asymptote $x = -\ln A$ and tending to 0 as $x \to -\infty$.

For $0 < y < L$ we write

$$-\ln(L-y) + \ln(y) = x + C$$

We then find

$$y = \frac{LAe^x}{Ae^x + 1}$$

where $A = e^C > 0$. This describes an increasing solution defined in $]-\infty, \infty[$, tending to 0 as $x \to -\infty$ and to L as $x \to \infty$.

Finally for $L < y$ we write

$$-\ln(y-L) + \ln(y) = x + C$$

We then find

$$y = \frac{LAe^x}{Ae^x - 1}$$

where $A = e^C > 0$. This describes a decreasing solution in the region $y > L$ defined in the interval $]-\ln A, \infty[$, with a vertical asymptote $x = -\ln A$ and tending to L as $x \to \infty$.

Note that solutions in all three strips can be expressed by the formula

$$y = \frac{Le^x}{e^x + B}$$

where $B = \pm A^{-1}$. We can even think of the constant solutions as corresponding to the exceptional values $B = 0$ and $B = \infty$.

2.1.2 *The Non-autonomous Case*

Now we consider the more general problem

$$y' = h(x)g(y)$$

We assume that $h(x)$ is continuous in an open interval $]a_1, a_2[$ and that $g(y)$ is continuously differentiable in an open interval $]b_1, b_2[$.

The problem is therefore posed in the open rectangle

$$D = \{(x, y) : a_1 < x < a_2,\ b_1 < y < b_2\}.$$

Again the zeros of $g(y)$ give rise to the constant solutions. All other solutions can be described by the general solution obtained by separating the variables

$$\int \frac{dy}{g(y)} = \int h(x)\,dx + C. \tag{2.7}$$

As before we must be careful about choosing the antiderivative of $1/g(y)$. Consider the case of a solution curve in the region $y_1 < y < y_2$ where y_1 and y_2 are consecutive zeros of $g(y)$. By the same argument as we used in the autonomous case the solution curve is trapped in the horizontal strip $y_1 < y < y_2$. Since $g(y) \neq 0$ in this interval we can find an antiderivative for $1/g(y)$ on it. Call it $G(y)$ and let $H(x)$

be an antiderivative for $h(x)$ on $]a_1, a_2[$. Let the Cauchy condition be $y(x_0) = y_0$ where $y_1 < y_0 < y_2$. Our solution is found by solving the equation

$$G(y) - G(y_0) = H(x) - H(x_0) \tag{2.8}$$

for y as a function of x.

We now give an argument to show that the solution of the last paragraph is extendable over all of $]a_1, a_2[$. The reasoning is the same as that used in the last section, but we give more details.

We fix the antiderivative, for example by letting

$$G(y) = \int_{y_0}^{y} \frac{dt}{g(t)}, \quad (y_1 < y < y_2) \tag{2.9}$$

Let's suppose that $g(y) > 0$ on the interval $]y_1, y_2[$. We claim that the integral in (2.9) is divergent to $-\infty$ at $y = y_1$ and divergent to $+\infty$ at $y = y_2$ (if $g(y) < 0$ then the plus and minus are reversed). To see this we note that, since $g(y)$ is differentiable and $g(y_1) = 0$, we can write $g(y) = g_1(y)(y - y_1)$, where $g_1(y)$ is continuous. Let $|g_1(y)| < k$ on the interval $]y_1, y_1 + \delta[$. Then we have

$$\left| \frac{1}{g(y)} \right| \geq \frac{1}{k} \frac{1}{y - y_1}$$

on $]y_1, y_1 + \delta[$. By comparison with the divergent integral

$$\int_{y_1}^{y_1+\delta} \frac{dt}{t - y_1}$$

we conclude that the integral in (2.9) is divergent at $y = y_1$. A similar argument holds at $y = y_2$.

Now we see that $G(y)$, which is either strictly increasing or strictly decreasing according to the sign of $g(y)$, maps the interval $]y_1, y_2[$ on to the interval $]-\infty, \infty[$ (look back at Fig. 2.1). Hence it has an inverse $G^{-1} :]-\infty, \infty[\to]y_1, y_2[$. Applying this to (2.8) we obtain the solution

$$y = G^{-1}(H(x) - H(x_0) + G(y_0))$$

for all x within the domain of $H(x)$, that is, on all of $]a_1, a_2[$.

A similar analysis can be given of solution curves in the region above the highest zero of $g(y)$ or below the lowest zero. The essential point is to study the convergence or divergence of $\int (1/g(y))\, dy$ as $y \to b_1, b_2$ and relate it to the continuation of the solution curves. It is left to the reader to carry this out.

2.1.3 Exercises

1. Solve the following equations and sketch the solution curves. A good sketch should indicate where the constant solutions are, which solution curves extend to the whole line, and which have a vertical asymptote.

   (a) $y' = y$
   (b) $y' = y^2$
   (c) $y' = y(1 - y^2)$
   (d) $y' = xy^2$
   (e) $y' = y \cos x$
   (f) $y' = y^2 \cos x$

2. Solve the equation

$$y' = \cos y$$

   and sketch the solution curves.
   Hint By the properties of $\cos y$ it is enough to find solution curves for the region $-\pi/2 < y < \pi/2$. All others can be obtained from these by translation and reflection.

3. A practical application. An object falling vertically from a small height (of the order of a few thousand metres) towards the ground experiences the acceleration of gravity g, which can be considered constant, and atmospheric drag. For small heights, over which we can consider the air pressure to be constant, the drag can be taken to be proportional to the square of the velocity v. This results in the equation

$$\frac{dv}{dt} = g - kv^2.$$

   The constant k depends on a number of factors, including the air pressure and the shape of the object. Show that if the object starts from rest, the velocity v increases and approaches a limit, the terminal velocity. Find the limit.

4. Another practical application. A biological population of magnitude y (think of the number of bacteria on a plate, or the weight of a population of fish etc.) grows exponentially when y is small, but for larger y experiences a growth retardation owing to competition for resources. The latter can be modelled as a drag term proportional to y^2. This leads to the model

$$\frac{dy}{dt} = ay - by^2$$

   where a and b are positive constants. Show that the population does not increase indefinitely but tends to a positive limit. Find the limit.

5. Find and sketch the solution curves of the equation

$$y' = \sqrt{1 - y^2}$$

in the domain $-1 < y < 1$. You should find that solution curves cannot be extended to the whole line $-\infty < x < \infty$, even though they lie between two zeros of $\sqrt{1 - y^2}$, because they reach 1 or -1 at a finite value of x. Why does this not disagree with the analysis of the text?

6. (a) Obtain a general solution for the equation

$$y' = \sqrt{\frac{1 - y^2}{1 - x^2}}$$

in the square region D of the plane defined by $-1 < x < 1, -1 < y < 1$.

(b) Show that if the graph $y = \phi(x)$ is a solution curve on the interval $]a, b[$ then the graph $y = -\phi(-x)$ is a solution curve on the interval $]-b, -a[$. In other words the set of solution curves is invariant under the transformation $(x, y) \mapsto (-x, -y)$.

(c) Show that one solution curve is given by $y = x$, for $-1 < x < 1$.

(d) Show that all solution curves in D, other than $y = x$, are arcs of ellipses (including arcs of circles) that are inscribed in D, and whose axes lie along the lines $x + y = 0$ and $x - y = 0$.

(e) Sketch the solution curves.

Hint The general solution may not give a transparent way to sketch the solution curves. Use instead the conclusion of item (d). Remember, when sketching the curves, that only one solution curve can pass through a given point. Your curves must not cross each other.

7. Let f be a C^1 function, defined on the whole line $\mathbb{R}$ and let $f(y) > 0$ for all y. Suppose also that f is periodic with positive period α. Let $\phi(x)$ be a solution of $y' = f(y)$.

(a) Show that the solution $\phi(x)$ can be extended to the whole line $\mathbb{R}$.

(b) Show that

$$\phi(x) + \alpha = \phi(x + L), \quad (x \in \mathbb{R})$$

where

$$L = \int_0^\alpha \frac{dy}{f(y)}$$

It follows that the whole solution curve $y = \phi(x)$ can be built using the arc over the interval $0 < x < L$, by repeatedly translating it horizontally by L, and vertically by α.

8. We consider the first order separable equation $y' = h(x)g(y)$, where h is continuous and periodic with period T, whilst g is continuously differentiable in $\mathbb{R}$.

   (a) Show that if h has mean value 0 then all solutions that are between two constant solutions are periodic, but if h has non-zero mean value then only constant solutions are periodic.
   Note The case of mean value 0 is exemplified by Exercise 1 (items (e) and (f)).
   (b) Consider now the case where $g(0) = 0$ and $g(y) > 0$ for $0 < y < \infty$. Suppose also that h has mean value 0. Is it the case that all solutions in the region $y > 0$ are periodic? Or only some? Draw some examples showing what may happen.

9. Some equations can be transformed into a separable equation by a change of variable. An important example, somewhat confusingly known as the homogeneous equation, is an equation of the form

$$y' = f\left(\frac{y}{x}\right)$$

   in which f is a function of one variable.

   (a) Show that this equation is separable in the variables x, u where $u = y/x$.
   (b) Try this on the equation

$$y' = \frac{x+y}{x-y}$$

   You should obtain an implicit description of the solution curves on reverting to the variables x, y.
   (c) Find a much nicer description of the solution curves in polar coordinates.

   Note The domain for this equation is the plane minus the line $y = x$, and a strict interpretation would limit it to one of the two connected components of this set. For another interpretation, perhaps more natural in view of the result of item (c), see Sect. 4.1, project A on exact equations.

10. A second order equation of the form

$$y'' = f(y)g(y')$$

   can be converted to a separable first order equation, and integrated, by writing

$$p = \frac{dy}{dx}$$

   and using y as a new independent variable.

(a) Show that the outcome is the separable equation

$$\frac{dp}{dy} = \frac{f(y)g(p)}{p}$$

From a solution $p = \phi(y)$ of this equation a solution of the original equation can be found by integrating the separable equation

$$\frac{dy}{dx} = \phi(y).$$

(b) Solve the equation

$$y'' = y\,(y')^2$$

by quadratures.

Note The topic of Sect. 2.3 is a special case of the equation studied in this exercise.

2.2 One-Parameter Groups of Symmetries

If the equation $y' = f(x, y)$ is not separable, the question arises as to whether one can find a transformation of variables $\xi = \phi_1(x, y)$, $\eta = \phi_2(x, y)$, such that the equation is separable in the coordinates (ξ, η). An example was Exercise 9 in the previous section. It was an early observation that if the equation is invariant under a one-parameter group of symmetries,[2] then finding suitable coordinates (ξ, η) is greatly facilitated.

We begin with some definitions, which may appear at first sight overly abstract. However, the examples we shall use to illustrate the theory are really quite simple.

Definition An equation

$$\frac{dy}{dx} = f(x, y)$$

is said to be *invariant under a change of variables*

$$\xi = \phi_1(x, y), \quad \eta = \phi_2(x, y)$$

[2] The observation is usually attributed to S. Lie, who, starting in 1871, developed the theory of what are known nowadays as Lie groups.

if, in the new variables, we obtain the equation

$$\frac{d\eta}{d\xi} = f(\xi, \eta),$$

that is, an equation with the same function f as before.

As a simple example we can consider an autonomous equation $dy/dx = f(y)$. The translation $\xi = x + c$, $\eta = y$ leads to the equation $d\eta/d\xi = f(\eta)$, the same as we started with, but in the new coordinates.

A change of variables corresponds to a diffeomorphism $\phi : \mathbb{R}^2 \to \mathbb{R}^2$, by which we mean a bijective differentiable mapping with differentiable inverse. More generally, we could have a diffeomorphism $\phi : D_1 \to D_2$ where D_1 and D_2 are open domains in $\mathbb{R}^2$. The equation $dy/dx = f(x, y)$ is invariant under the diffeomorphism ϕ when it is invariant under the change of variables $\xi = \phi_1(x, y)$, $\eta = \phi_2(x, y)$, the functions ϕ_1 and ϕ_2 being the coordinates (components) of ϕ.

Definition By a *one-parameter group of diffeomorphisms* $\phi_t : D \to D$, is meant a family of diffeomorphisms $(\phi_t)_{t\in\mathbb{R}}$, parametrised by $t \in \mathbb{R}$, that satisfies the group property

$$\phi_s \circ \phi_t = \phi_{s+t}, \qquad (s, t \in \mathbb{R})$$
$$\phi_0 = \text{identity}$$

We also require $\phi_t(x, y)$ to be a differentiable function of t for each (x, y). The differentiable curve $t \mapsto \phi_t(x, y)$ is called the *orbit* of the point (x, y).

In the language of group theory, we are describing an action of the additive group $(\mathbb{R}, +)$ on the domain D.

Definition An equation $dy/dx = f(x, y)$ is said to be invariant under the group $(\phi_t)_{t\in\mathbb{R}}$ when it is invariant under each diffeomorphism ϕ_t.

Sometimes it is convenient to use the multiplicative group $(\mathbb{R}_+, \times)$ of positive real numbers, instead of the additive group of all real numbers. Then we require the family $(\phi_t)_{t\in\mathbb{R}_+}$ of diffeomorphisms to satisfy

$$\phi_s \circ \phi_t = \phi_{st}, \qquad (s, t \in \mathbb{R}_+)$$
$$\phi_1 = \text{identity}$$

Of course the two groups are really the same, since the exponential function $t \mapsto e^t$, mapping $\mathbb{R}$ on to $\mathbb{R}_+$, is an isomorphism between them. So which one we use is a matter of convenience.

Examples of Groups and Invariant Equations

(1) $\phi_t(x, y) = (x + t, y), \quad (t \in \mathbb{R})$.

Here we have the group of translations in the x-variable. It is easy to see that the equation $dy/dx = f(x, y)$ is invariant under this group precisely when f is independent of x, so that such an equation is already separable.

Closely related (so closely that we give it the same number) is the group

(1′) $\phi_t(x, y) = (tx, y), \quad (t \in \mathbb{R}_+)$.

This is the group of dilatations in the x-coordinate, written multiplicatively. Now an equation $dy/dx = f(x, y)$ is invariant if and only if $f(x, y) = x^{-1} f(1, y)$, and the equation is already separable.

(2) $\phi_t(x, y) = (x, y + t), \quad (t \in \mathbb{R})$.

Here we have the group of translations in the y-variable. It is easy to see that the equation $dy/dx = f(x, y)$ is invariant under this group precisely when f is independent of y, so that such an equation is already separable.

Closely related is the group

(2′) $\phi_t(x, y) = (x, ty), \quad (t \in \mathbb{R}_+)$.

This is the group of dilatations in the y-coordinate, written multiplicatively. Now an equation $dy/dx = f(x, y)$ is invariant if and only if $f(x, y) = yf(x, 1)$, and the equation is already separable.

(3) $\phi_t(x, y) = (tx, ty), \quad (t \in \mathbb{R}_+)$.

Here we have the group of dilatations, defined in the plane domain $D = \mathbb{R}^2 \setminus \{O\}$, and written multiplicatively. The equation $dy/dx = f(x, y)$ is invariant under this group precisely when $f(tx, ty) = f(x, y)$ for all $(x, y) \in D$ and $t > 0$. Equivalently, f is a function of the ratio y/x. An example is the equation

$$\frac{dy}{dx} = \frac{x}{x + y}$$

(4) $\phi_t(x, y) = (t^3 x, ty), \quad (t \in \mathbb{R}_+)$.

Here we have a group of *weighted dilatations*, defined in the same domain D as in example 3. The equation $dy/dx = f(x, y)$ is invariant under this group precisely when $f(t^3 x, ty) = t^{-2} f(x, y)$ for all $(x, y) \in D$ and $t > 0$. An example is the equation

$$\frac{dy}{dx} = \frac{1}{y^2} + \frac{y}{x}.$$

(5) $\phi_t(x, y) = (x \cos t - y \sin t, x \sin t + y \cos t), \quad (t \in \mathbb{R})$.

Here we have the group of *rotations*, defined in the same domain D as in example 3. The equation $dy/dx = f(x, y)$ is invariant under this group precisely when

$$\frac{f(\phi_t(x, y))\cos t - \sin t}{\cos t + f(\phi_t(x, y))\sin t} = f(x, y), \quad \big((x, y) \in D,\ t \in \mathbb{R}\big)$$

An example is the equation

$$\frac{dy}{dx} = \frac{x^3 + xy^2 + y}{x - x^2y - y^3}.$$

Exercise For each example in the above list, verify the condition stated for the equation $dy/dx = f(x, y)$ to be invariant under the given group. Verify also that the equation given as an example is invariant.

For examples 1, 1′, 2 and 2′ the equation is already separable. How does the group help us to solve the equation in the other cases? We consider the *orbit of a point* (x, y), that is, the curve $t \mapsto \phi_t(x, y)$, which passes through (x, y). The orbits partition the domain of the equation in the plane into disjoint differentiable curves. Through a point (x_0, y_0) we draw a curve C, parametrised as $\eta \mapsto (\gamma_1(\eta), \gamma_2(\eta))$, such that C is transversal to the orbit through (x_0, y_0), meaning that the tangent to C at (x_0, y_0) and the tangent to the orbit at (x_0, y_0) are independent vectors. It could be convenient, though it is not necessary, to arrange for $\eta = 0$ at (x_0, y_0). If (x, y) is in a neighbourhood of the point (x_0, y_0) we can follow the orbit through (x, y) until it crosses the curve C. In other words we define η and ξ such that $(x, y) = \phi_\xi(\gamma_1(\eta), \gamma_2(\eta))$. Then $\xi = 0$ on C and $\eta = 0$ on the orbit through (x_0, y_0). We use (ξ, η) as new coordinates in a neighbourhood of (x_0, y_0). This is illustrated in Fig. 2.3.

In the coordinates (ξ, η) the action of the group is nothing other than translation $(\xi, \eta) \mapsto (\xi + t, \eta)$ (or, if the group is written multiplicatively, it is dilatation in

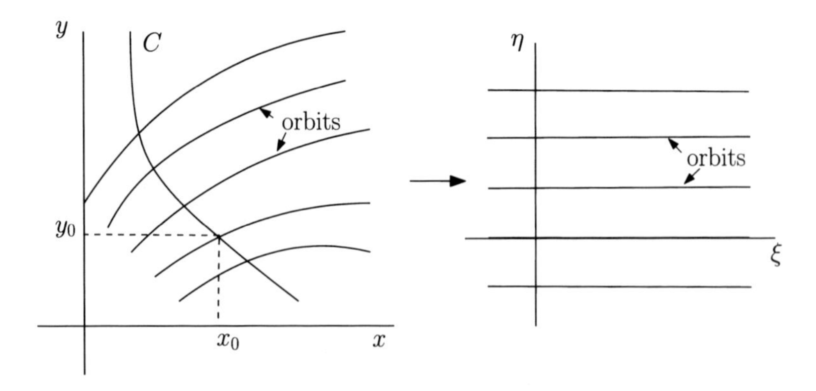

Fig. 2.3 Using the group to find new variables

the ξ-coordinate $(\xi, \eta) \mapsto (t\xi, \eta)$). Therefore, in these coordinates the differential equation is separable, as in examples 1 and 1′.

In practice, it is often convenient to implement this by finding a function $h(x, y)$ that is invariant under the action of the group, that is, h satisfies

$$h(\phi_t(x, y)) = h(x, y), \quad (t \in \mathbb{R} \text{ or } t \in \mathbb{R}_+ \text{ as appropriate}).$$

The function h is therefore constant on orbits. Now if in some region we have $\partial h/\partial y \neq 0$, then lines in the region of the form $x =$ constant are transversal to the orbits. Choose one such line as C, for example $x = x_0$ and parametrise it by the values of h along it. That is, we take $\eta = h(x, y)$. Now we can define ξ so that $(\phi_{-\xi})_1(x, y) = x_0$, where $(\phi_{-\xi})_1$ is the first component of $\phi_{-\xi}$, or, if the group is written multiplicatively, $(\phi_{1/\xi})_1(x, y) = x_0$. In the coordinates (ξ, η) the equation is separable.

Example for Finding New Coordinates

Consider the equation

$$y' = \frac{x + y}{x - y}$$

It is simple to see that it is invariant under the multiplicative group of dilatations $\phi_t(x, y) = (tx, ty)$. An obvious invariant function of this group is y/x. So one of the new coordinates is $\eta := y/x$. We can take C as the line $x = 1$, as illustrated in Fig. 2.4. The orbit through (x, y) crosses C at $(1, y/x)$, and $(x, y) = \phi_x(1, y/x)$. So we set $\xi := x$. Showing that the equation separates in the new coordinates was part of 2.2 Exercise 9. Actually, this equation is also invariant under the rotation group; this should be plausible to those readers who have worked through 2.2 Exercise 9. See Exercise 3.

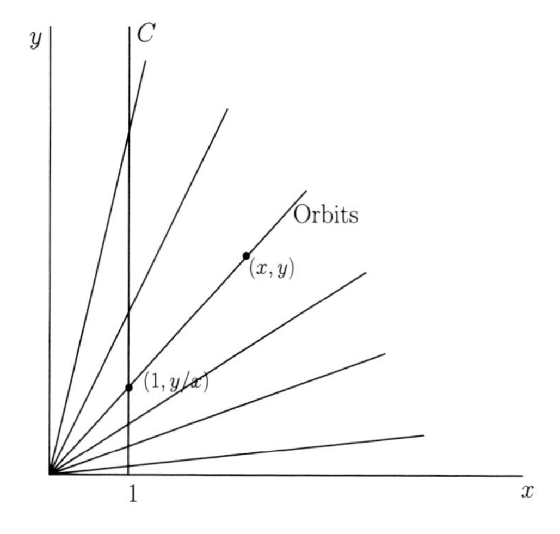

Fig. 2.4 Orbits of the dilatation group

We shall return to the subject of symmetry groups of equations in Sect. 4.1 Project A, where we look at the notion of integrating factors, and how symmetry groups can help us to find them.

2.2.1 Exercises

1. For examples 3, 4 and 5 (of groups and invariant equations) find coordinates (ξ, η), in which η is an invariant function of the group and the variables can be separated. Proceed to separate the variables and solve the example equations as far as is practically possible.
   Hint For example 3 you can use $\eta = y/x$, for example 4 use $\eta = y^3/x$ and for example 5 use $\eta = \sqrt{x^2 + y^2}$ (polar radial coordinate). Often an implicit relation between x and y is as far as one can get using analytical methods.
2. Show that the equation
$$\frac{dy}{dx} = (xy + 1)y^2$$
   is invariant under the group $\phi_t(x, y) = (t^{-1}x, ty)$. Find coordinates in which the equation separates, and hence obtain an implicit equation for its solution curves in the form $F(x, y) = C$, where C is an arbitrary constant.
3. Verify that the equation
$$\frac{dy}{dx} = \frac{x + y}{x - y}$$
   is invariant under the rotation group. Show that the equation separates in plane polar coordinates r, θ (as it must do by the construction of new coordinates described in the text) and express its solution curves in polar coordinates.
4. Invariance of an equation $dy/dx = f(x, y)$ under a one-parameter group is equivalent to requiring f to satisfy a certain first-order partial differential equation.

   (a) For the groups listed under 1, 1′, 2 and 2′ show that the equations are respectively
$$\frac{\partial f}{\partial x} = 0, \qquad x\frac{\partial f}{\partial x} + f = 0, \qquad \frac{\partial f}{\partial y} = 0, \qquad y\frac{\partial f}{\partial y} - f = 0.$$

(b) Obtain the following equivalent partial differential equations in examples 3, 4 and 5:

$$x\frac{\partial f}{\partial x} + y\frac{\partial f}{\partial y} = 0, \qquad \text{(case 3)}$$

$$3x\frac{\partial f}{\partial x} + y\frac{\partial f}{\partial y} = -2f, \qquad \text{(case 4)}$$

$$-y\frac{\partial f}{\partial x} + x\frac{\partial f}{\partial y} = 1 + f^2, \qquad \text{(case 5)}$$

Hint In each case, differentiate the defining property with respect to t. For example, in case 4 it is $t^2 f(t^3x, ty) = f(x, y)$.

2.3 Newton's Equation

In this section we study another class of equations that can be solved by quadratures. Newton's equation is the second order equation

$$\frac{d^2x}{dt^2} = F(x).$$

It is so called because we can think of $F(x)$ as a force field, x as displacement along a straight line and d^2x/dt^2 as acceleration. We have here the equation of motion of a particle in one dimension, or, as physicists would say, a mechanical system with one degree of freedom. Newton's equation can be integrated, essentially by reducing it to a family of separable first order equations, parametrised by a constant called the energy. This, together with its great historical importance and practical value, amply justifies including it under the heading "essential".

To reflect the physical picture we adopt a different notation in this section from that used previously. The independent variable is t, thought of as time, and henceforth referred to as time. The dependent variable is x, thought of as displacement along a line with coordinate x. Derivatives with respect to t will be denoted by dashes, thus: x' (velocity), x'' (acceleration) and so on.

Newton's equation is equivalent to a system of first order equations in a plane with coordinates x and y, namely:

$$x' = y, \quad y' = F(x)$$

The quantity y is called the velocity. A solution $(x(t), y(t))$ of this system is usually viewed as a parametrised curve in the (x, y)-plane, and is known as a *phase curve*, or even, emphasising the physical picture, a *motion of the system*. Of course if $(x(t), y(t))$ is a phase curve then the first coordinate $x(t)$ is a solution of Newton's equation and might describe the *physical motion of a particle*, sometimes leading to a slight ambiguity. Conversely if $x(t)$ is a solution of Newton's equation then the pair $(x(t), x'(t))$ is a phase curve.

The (x, y)-plane is also known as the *phase plane*. The vector field $(y, F(x))$ in the phase plane is called the *phase velocity*. It is the velocity at the point (x, y) along the phase curve through (x, y). It is a vector and should not be confused with the physical velocity x', which is a scalar. Clearly all motions in the region $y > 0$ go to the right (that is, in the direction of increasing x), whilst all motions in the region $y < 0$ go to the left.

We assume that F is at least C^1. Then, by the existence theorem to be proved in Chap. 4, there exists, for each (x_0, y_0) in the phase plane, a unique phase curve $(x(t), y(t))$ defined for some interval $-h < t < h$, such that $x(0) = x_0$ and $y(0) = y_0$. The existence of this phase curve will become clear without the help of the existence theorem, since we will essentially obtain it by quadratures.

We introduce the potential energy $U(x)$, defined to be an antiderivative of $-F(x)$,

$$U(x) = -\int F(x)\,dx.$$

The potential energy can be altered by the addition of a constant. It may therefore be important to specify clearly what is being used as the potential energy in any given application.

Proposition 2.1 *The quantity*

$$E(x, y) = U(x) + \frac{1}{2}y^2$$

(called the energy) is a constant of the motion.

The meaning of *constant of the motion* is this: for every phase curve $(x(t), y(t))$ the quantity $E\big(x(t), y(t)\big)$ is constant, that is, independent of t.

Proof It is enough to observe that the gradient of $E(x, y)$ is the vector $(-F(x), y)$, which is orthogonal to the phase velocity $(y, F(x))$, and the latter is tangent to the phase curve through (x, y). □

It follows that every phase curve lies in a level set of the function E. Consider therefore the level set

$$H_c = E^{-1}(c).$$

The set H_c will in general include a number of distinct phase curves, each of which may be called a motion at the energy level c. Because E possesses reflexion symmetry, that is $E(x, y) = E(x, -y)$, we see that H_c is symmetrical with respect to the x-axis. Moreover the phase velocities at (x, y) and at $(x, -y)$ have the same

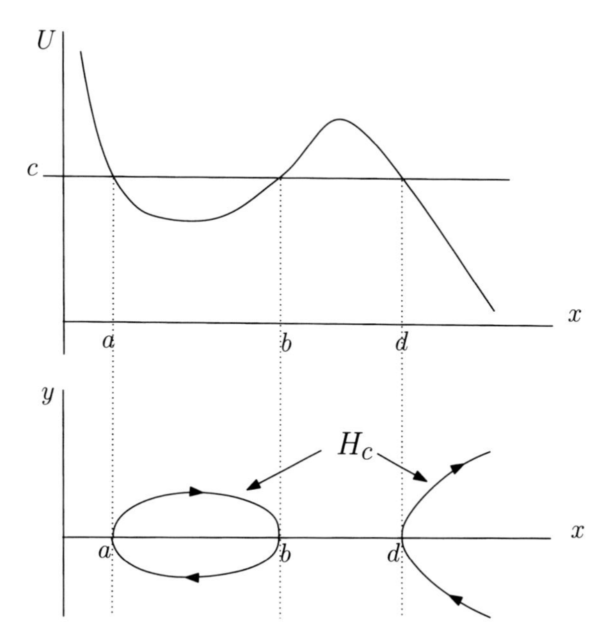

Fig. 2.5 Sketching the phase curves at energy level c: case 1

length, so that, given two points (x_1, y_1) and (x_2, y_2) in the same phase curve in the upper half plane, the time of passage from (x_1, y_1) to (x_2, y_2) is the same as the time of passage from $(x_2, -y_2)$ to $(x_1, -y_1)$ in the lower half plane.

If $U(x) + \frac{1}{2}y^2 = c$ then $U(x) \leq c$. The projection of the set H_c to the x-axis is therefore the set $\{x : U(x) \leq c\}$. We say that c is a *regular level* if $U'(x) \neq 0$ for every x such that $U(x) = c$.

The illustrations in Figs. 2.5 and 2.6 should be helpful in the ensuing discussion of regular levels.

2.3.1 Motion in a Regular Level Set

Proposition 2.2 *If c is a regular level and H_c non-empty then H_c is a differentiable curve (possibly disconnected) that does not intersect itself.*

Proof Let $(x_0, y_0) \in H_c$. If $y_0 \neq 0$ then $\frac{\partial E}{\partial y}(x_0, y_0) = y_0 \neq 0$. If $y_0 = 0$ then $U(x_0) = 0$ and so $\frac{\partial E}{\partial x}(x_0, y_0) = U'(x_0) \neq 0$. The conclusion follows by the implicit function theorem. More precisely, in a neighbourhood of each of its points either H_c is a graph of the form $y = Y(x)$, or else it is a graph of the form $x = X(y)$, the functions $X(y)$ and $Y(x)$ being C^1. □

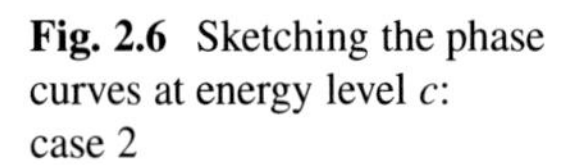

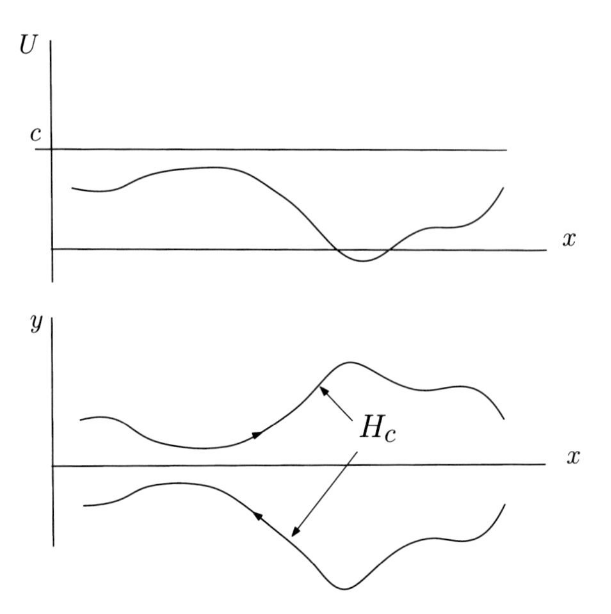

Fig. 2.6 Sketching the phase curves at energy level c: case 2

We are going to analyse motions that take place in a regular level set H_c. It turns out that it is easy to sketch the phase curves without determining exactly how x and x' depend on t. There are two principal cases:

Case 1. The fibre $U^{-1}(c)$ is non-empty.

In this case the set $\{x : U(x) < c\}$ is also non-empty (because, c being a regular level, if $U(x) = c$ then $U'(x) \neq 0$). The set $\{x : U(x) < c\}$ is an open subset of the real line, and as such, it is a countable disjoint union of open intervals. Because c is a regular level no two of these intervals can share an endpoint. Let $]a, b[$ be one of these intervals, supposed bounded. We have $U(a) = U(b) = c$. That part of the set H_c for which $a \leq x \leq b$ consists of a differentiable curve, symmetrical with respect the x-axis, that intersects the x-axis at the points a and b. It is diffeomorphic to a circle. The illustration in Fig. 2.5 may be helpful.

From the equation $E(x, y) = c$ we obtain, for that part of the curve for which $y > 0$, the differential equation

$$x' = \sqrt{2c - 2U(x)},$$

and for that part for which $y < 0$,

$$x' = -\sqrt{2c - 2U(x)}$$

Both these equations can be solved, provided x remains between a and b, by separating the variables. Choose some point on the upper part of the curve as the initial point, say with $x = x_0$, where $a < x_0 < b$. Then on the upper curve we have

$$\int_{x_0}^{x} \frac{d\xi}{\sqrt{2c - 2U(\xi)}} = t, \tag{2.10}$$

and this relation can be inverted to express x as a function of t. This works as long as x remains between a and b. The time of transit from a to b is then

$$T = \int_{a}^{b} \frac{d\xi}{\sqrt{2c - 2U(\xi)}}. \tag{2.11}$$

Exercise Show that the improper integral converges at both ends.

Hint $U'(a) \neq 0$, $U'(b) \neq 0$.

The coordinate x reaches b at time t_1 given by

$$t_1 = \int_{x_0}^{b} \frac{d\xi}{\sqrt{2c - 2U(\xi)}}.$$

The phase point (x, y) then passes to the lower plane and x returns to a. The dependence of x on t in the lower plane is then the mirror image of what it was in the upper plane, that is,

$$x(t) = x(2t_1 - t),$$

and continues until x reaches a and the phase point (x, y) passes back to the upper plane.

The result of this is that x is a periodic function of t with period $2T$. Since $y = x'$ it is clear that $y(t)$ is also a periodic function with the same period.

The set $\{x : U(x) < c\}$ may have a connected component of the form $]d, \infty[$, where d is a finite number (as suggested by the illustration in Fig. 2.5). The motion clearly proceeds from the lower plane, moving to the left until x reaches d, and passes to the upper plane, moving to the right. Again it satisfies (2.10) in the upper plane (taking some initial x_0 to the right of d), and x escapes to infinity in the time

$$\int_{x_0}^{\infty} \frac{d\xi}{\sqrt{2c - 2U(\xi)}}$$

a quantity that may be finite or infinite depending on the function $U(x)$. A similar analysis can be made for a connected component of the form $]-\infty, d[$.

Exercise Show that if $U(x)$ is bounded below, the time of escape to infinity is infinite.

Case 2. The fibre $U^{-1}(c)$ is empty but H_c is non-empty.

This happens if $U(x) < c$ for all x. The level set H_c consists of the two curves

$$y = \pm\sqrt{2c - 2U(x)},$$

both defined for all x (Fig. 2.6). The motion in the upper plane is again given by inverting (2.10).

2.3.2 Critical Points

A number c is said to be a critical value if there exists x_0 such that $U(x_0) = c$ and $U'(x_0) = 0$. Clearly we then have a solution with energy c given by $x(t) = x_0$, a constant solution. Physically this describes an equilibrium point.

We will analyse the structure of the phase curves near $(x_0, 0)$, at energy levels near to c, in the important case that $U''(x_0) \neq 0$. In this case we say that x_0 is a *non-degenerate critical point* of $U(x)$. We assume that U is a C^∞ function. Then we can write, using Taylor's theorem:

$$U(x) = c + \int_{x_0}^{x} (x - \xi)U''(\xi)\, d\xi,$$

and using the substitution $\xi = x_0 + \sigma(x - x_0)$ in the integral we find

$$U(x) = c + (x - x_0)^2 \int_0^1 (1 - \sigma)U''\big(x_0 + \sigma(x - x_0)\big)\, d\sigma,$$

that is,

$$U(x) = c + \tfrac{1}{2}(x - x_0)^2 G(x)$$

where $G(x)$ is a C^∞ function that satisfies $G(x_0) = U''(x_0)$.

We now distinguish two cases:

Case A. $U''(x_0) > 0$.

We can introduce new coordinates (ξ, y) in the phase plane in a neighbourhood of $(x_0, 0)$ by setting

$$\xi = (x - x_0)\sqrt{G(x)}.$$

Phase curves near the point $x = x_0$, $y = 0$, that is near $\xi = 0$, $y = 0$ using the new coordinates, on the energy level $c + \varepsilon$, are given by the equation

$$\tfrac{1}{2}\xi^2 + \tfrac{1}{2}y^2 = \varepsilon$$

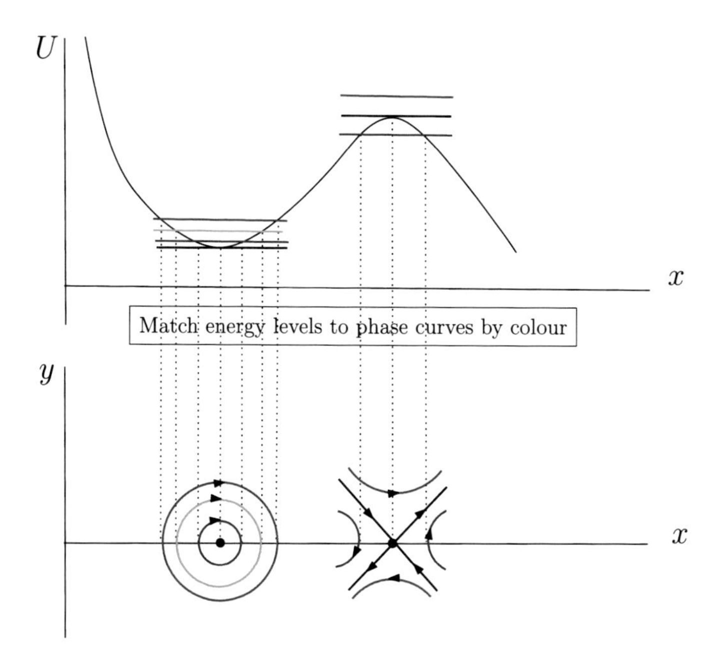

Fig. 2.7 Motion near critical points of U

where ε is positive (because $(x_0, 0)$ is a local minimum point for energy) and sufficiently small. They are concentric circles in the new coordinates and represent oscillations with amplitude $\sqrt{2\varepsilon}$. The motion starting at a point close to the equilibrium point $(x_0, 0)$, remains close to it. For this reason the equilibrium point is said to be *stable*. The motion near the critical point is illustrated in Fig. 2.7.

Case B. $U''(x_0) < 0$.

We can introduce new coordinates (ξ, y) in the phase plane in a neighbourhood of $(x_0, 0)$ by setting

$$\xi = (x - x_0)\sqrt{-G(x)}.$$

Phase curves near the point $x = x_0$, $y = 0$, that is near $\xi = 0$, $y = 0$ using the new coordinates, on the energy levels $c \pm \varepsilon$, are given by the equation

$$-\frac{1}{2}\xi^2 + \frac{1}{2}y^2 = \pm\varepsilon$$

where ε is positive and sufficiently small. In the new coordinates they are a family of concentric hyperbolae with common asymptotes, together with the asymptotes themselves, the two lines $\xi \pm y = 0$.

The motion starting at a point close to the equilibrium point $(x_0, 0)$ diverges from it, unless it begins precisely on the asymptote $\xi + y = 0$. For this reason the equilibrium point is said to be *unstable*. This is illustrated in Fig. 2.7.

The notion of stability for equilibrium points (or constant solutions) of differential equations will be studied more generally in Sect. 6.3.

Small Oscillations

In case A, near the critical point x_0, the system executes small oscillations, following the closed phase curves that surround $(x_0, 0)$ in the phase plane (Fig. 2.7). The period on a level with energy $c + \varepsilon$ is given by

$$L_\varepsilon = 2\int_a^b \frac{dx}{\sqrt{2(c+\varepsilon) - 2U(x)}}$$

where a and b are the intersection points of the level curve with energy $c + \varepsilon$ with the x-axis nearest to x_0, below it and above it respectively. Using the new coordinate $\xi = (x - x_0)\sqrt{G(x)}$ we rewrite this as

$$L_\varepsilon = 2\int_{-\sqrt{2\varepsilon}}^{\sqrt{2\varepsilon}} \frac{J(\xi)}{\sqrt{2\varepsilon - \xi^2}}\, d\xi,$$

where $J(\xi) = dx/d\xi$. A further substitution, setting $\xi = \sqrt{2\varepsilon}\,\sigma$ leads to

$$L_\varepsilon = 2\int_{-1}^{1} \frac{J(\sqrt{2\varepsilon}\,\sigma)}{\sqrt{1-\sigma^2}}\, d\sigma.$$

The limit $\lim_{\varepsilon\to 0} L_\varepsilon$ is called the *period of small oscillations at the equilibrium point* x_0. We can easily compute this. Setting $J_0 = J(0)$ we have:

$$\lim_{\varepsilon\to 0} L_\varepsilon = 2J_0 \int_{-1}^{1} \frac{1}{\sqrt{1-\sigma^2}}\, d\sigma = 2\pi J_0 = \frac{2\pi}{\sqrt{G(x_0)}} = \frac{2\pi}{\sqrt{U''(x_0)}}.$$

In physical applications we can expect that $2\pi J_0$ provides a reasonable approximation to the period of oscillation L_ε on phase curves near to the equilibrium point, and this expectation is reinforced by a heuristic calculation. We expand $J(\xi)$ in powers of ξ

$$J(\xi) = J_0 + J_1\xi + J_2\xi^2 + \cdots$$

where

$$J_k = \frac{1}{k!}\frac{d^{k+1}x}{d\xi^{k+1}}\bigg|_{\xi=0}.$$

We then have

$$L_\varepsilon = 2\int_{-1}^{1} \frac{1}{\sqrt{1-\sigma^2}} \left(J_0 + \sqrt{2\varepsilon} J_1 \sigma + 2\varepsilon J_2 \sigma^2 + \cdots \right) d\sigma.$$

Since

$$\int_{-1}^{1} \frac{\sigma}{\sqrt{1-\sigma^2}}\, d\sigma = 0$$

we can conclude by this non-rigorous argument that the limit

$$\lim_{\varepsilon \to 0+} \varepsilon^{-1} (L_\varepsilon - 2\pi J_0)$$

exists and is a finite number, whilst

$$\lim_{\varepsilon \to 0+} \varepsilon^{-1/2} (L_\varepsilon - 2\pi J_0) = 0.$$

Exercise Prove these limit formulas rigorously by replacing the power series for $J(\xi)$ by a Taylor expansion with remainder. Show that the first limit is $2\pi J_2$.

These conclusions are significant because ξ oscillates between $-\sqrt{2\varepsilon}$ and $+\sqrt{2\varepsilon}$. By the second limit the error involved in approximating L_ε by $2\pi J_0$ is vanishingly small compared to the amplitude when the amplitude is small. This is why the usual formula, given in the elementary textbooks of physics, for the period of a simple pendulum, which is actually the period of small oscillations, is practical even for moderate amplitudes (see Exercises 5, 6 and 7 for an exact treatment of the pendulum).

2.3.3 Exercises

1. A simple case of Newton's equation that has untold applications is the equation

$$\frac{d^2x}{dt^2} = -kx$$

where k is a positive constant. We take $U(x) = \frac{1}{2}kx^2$. The equation describes motion in one dimension under an attractive force towards the origin that is proportional to the distance. The motion is known as simple harmonic motion and a system subject to this law is known as a simple harmonic oscillator. Its great importance lies in the fact that real systems undergoing oscillations of small amplitude can usually be regarded to a good approximation as simple harmonic oscillators.

(a) Find the solution that satisfies $x(0) = a$, $x'(0) = 0$. Show that it is periodic if $a \neq 0$, find its period and note that it is independent of a.
(b) Find the most general solution that has energy c, where $c \geq 0$.

2. Consider the problem

$$\frac{d^2x}{dt^2} = -\frac{a}{x^2}, \qquad (0 < x < \infty)$$

where a is a positive constant. We take $U(x) := -a/x$. When equipped with the initial conditions $x(R) = 0$, $x'(R) = v$ the equation can describe a rocket shot vertically with speed v from the surface of a planet of radius R, and subject to an inverse square law of attraction to the centre.

(a) Sketch phase curves in the plane with coordinates x, x', having energies

$$E = -2, \quad -1, \quad 0, \quad 1, \quad 2.$$

(b) Find a formula for the least initial speed v for which the rocket reaches infinity (the escape velocity).

3. Consider the equation

$$\frac{d^2x}{dt^2} = -x + x^3.$$

(a) Sketch phase curves in the (x, x')-plane showing the three equilibrium points, closed phase curves that encircle one equilibrium point only, and others that encircle all three.
(b) Find the period of small oscillations around the two minima of the potential energy.

4. Sketch phase curves of Newton's equation for the potential shown in the graph in the accompanying drawing:

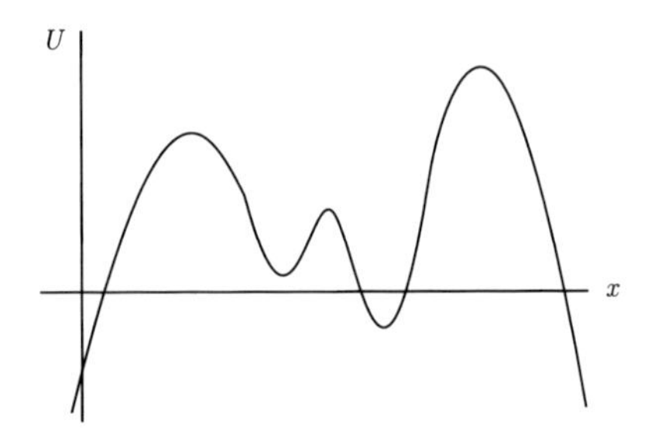

5. The equation of motion of a frictionless simple pendulum of length ℓ in a gravitational field of strength g, in terms of an angular coordinate θ, is

$$\frac{d^2\theta}{dt^2} = -\frac{g}{\ell}\sin\theta.$$

We take $U(\theta) = (g/\ell)(1 - \cos\theta)$; this choice has the advantage that the minimum of U is 0.

(a) Draw some phase curves in the plane with coordinates θ, θ', having energies

$$E = 0, \quad \frac{g}{2\ell}, \quad \frac{g}{\ell}, \quad \frac{3g}{2\ell}, \quad \frac{2g}{\ell}, \quad \frac{3g}{\ell}.$$

(b) Let $0 < \alpha < \pi$. Show that the solution with initial values $\theta = \alpha$, $\theta' = 0$ oscillates between α and $-\alpha$. Obtain the formula

$$T_\alpha = \sqrt{\frac{2\ell}{g}} \int_{-\alpha}^{\alpha} \frac{d\theta}{\sqrt{\cos\theta - \cos\alpha}}$$

for its period.

(c) Find the period of small oscillations, that is, the limit $\lim_{\alpha\to 0+} T_\alpha$.

Hint Finding the limit $\lim_{\alpha\to 0+} T_\alpha$ using the integral in item (b) is a bit tricky. It is easier to use the formula obtained in the text under 'Small oscillations'. If you want to use the integral it is best to begin with a change of variable. Some transformations of the integral have proved important; see Exercise 7.

6. Newton's equation is a source of some important periodic functions of analysis. The equation

$$\frac{d^2x}{dt^2} = 2k^2x^3 - (1 + k^2)x$$

gives rise to the Jacobi elliptic function $\mathrm{sn}_k(t)$. Here k is a constant in the range $0 \le k < 1$, called the modulus in elliptic function theory. When $k = 0$ we obtain the circular functions.

(a) The function $\mathrm{sn}_k(t)$ can be defined as the solution that satisfies the initial conditions $x(0) = 0$, $x'(0) = 1$. Show that $\mathrm{sn}_k(t)$ is periodic with period $2L$, where

$$L = \int_{-1}^{1} \frac{dx}{\sqrt{(1 - x^2)(1 - k^2x^2)}}.$$

Hint If we take $U(x) = -\frac{1}{2}k^2x^4 + \frac{1}{2}(1+k^2)x^2$ then the solution in question is at energy level $\frac{1}{2}$. If the corresponding phase curve crosses the x-axis it must be at a root of $U(x) = \frac{1}{2}$.

(b) Show that the maximum and minimum of $\mathrm{sn}_k(t)$ are, respectively, 1 and -1, and that $\mathrm{sn}_k(t)$ satisfies

$$\mathrm{sn}_k(t+L) = -\mathrm{sn}_k(t), \quad \mathrm{sn}_k(L-t) = \mathrm{sn}_k(t).$$

(c) Show that $w(t) := \mathrm{sn}_k(t)$ satisfies

$$w'^2 = (1-w^2)(1-k^2w^2).$$

7. The equation of motion of the simple pendulum given in Exercise 5 can be integrated using elliptic functions defined in Exercise 6.

(a) For a motion with amplitude α show that

$$\theta'^2 = \frac{4g}{\ell}\left(\sin^2\frac{\alpha}{2} - \sin^2\frac{\theta}{2}\right)$$

(b) Let

$$w = \frac{\sin\frac{1}{2}\theta}{\sin\frac{1}{2}\alpha}$$

and show that w satisfies

$$w'^2 = p^2(1-w^2)(1-k^2w^2)$$

where

$$k = \sin\tfrac{1}{2}\alpha, \quad p = \sqrt{\frac{g}{\ell}}.$$

(c) Obtain the most general solution with amplitude α, in terms of the Jacobi elliptic function sn_k (defined in Exercise 6), in the form

$$\theta = 2\arcsin\big(k\,\mathrm{sn}_k(pt+C)\big)$$

where C is an arbitrary constant.

(d) Show that the period of oscillation with amplitude α is

$$T_\alpha = 2\sqrt{\frac{\ell}{g}} \int_{-1}^{1} \frac{dw}{\sqrt{(1-w^2)(1-k^2w^2)}}$$

(e) Let $w = \sin\phi$ in the integral of the previous item and obtain

$$T_\alpha = 2\sqrt{\frac{\ell}{g}} \int_{-\frac{\pi}{2}}^{\frac{\pi}{2}} \frac{d\phi}{\sqrt{1-k^2\sin^2\phi}}$$

This is often the most convenient formula for studying the dependence of T_α on α.

Note The integral is one of Legendre's three standard elliptic integrals.

(f) Obtain the power series expansion

$$T_\alpha = 2\pi\sqrt{\frac{\ell}{g}} \sum_{n=0}^{\infty} c_n k^{2n}, \quad (0 \le k < 1)$$

where

$$c_0 = 1, \quad c_n = \left(\frac{2n-1}{2n}\,\frac{2n-3}{2n-2}\cdots\frac{1}{2}\right)^2.$$

2.4 Motion in a Central Force Field

The problem of motion in a central force field reduces to a one-parameter family of Newton's equations. Therefore, it can be solved by quadratures. This is the famous *one body problem*, which has two degrees of freedom, meaning that two spatial coordinates are needed to state it since the motion takes place in a plane. The case when the force field obeys an inverse square law is the famous Kepler problem, the solution of which by Newton established calculus as the fundamental mathematical tool for studying problems of motion in the physical world.

We saw in the previous section that problems with one degree of freedom can be solved by quadratures. The one body problem is one of only a small number of problems of classical mechanics that can be so solved once we allow more that one degree of freedom.[3] This section is structured like the projects that follow the exercise sections, but with a higher ratio of text to exercises.

[3] The two body problem can be reduced to the one body problem. Therefore, it too can be solved by quadratures. The real difficulties begin with the three body problem.

Fig. 2.8 The basis $(\mathbf{e}_r, \mathbf{e}_\theta)$

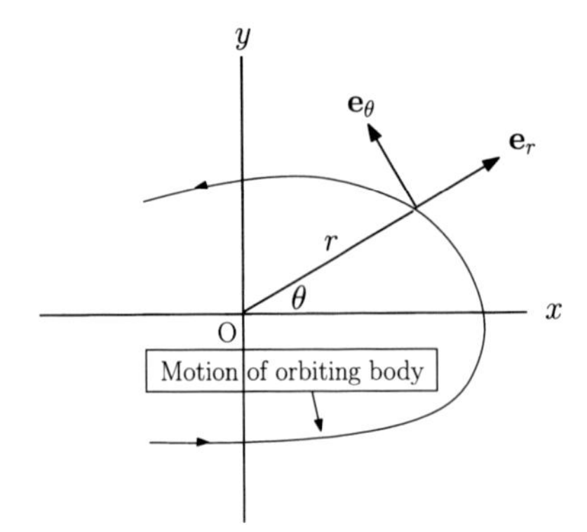

An acquaintance with multivariate calculus is required. We will use plane polar coordinates, the corresponding orthonormal frame for vectors and the expression of velocity and acceleration for a motion described in plane polar coordinates and referred to this frame. We denote by x and y the usual cartesian coordinates, and by r and θ the usual polar coordinates of the plane in which the motion takes place. The orthonormal frame for vectors is given by

$$\mathbf{e}_r = \left(\frac{x}{r}, \frac{y}{r}\right), \quad \mathbf{e}_\theta = \left(-\frac{y}{r}, \frac{x}{r}\right)$$

See the illustration in Fig. 2.8.

We consider a force field in the plane, which is a function of r alone and is directed along the radius. The force (per unit mass of the orbiting body) at a point (r, θ) is the vector $-F(r)\mathbf{e}_r$. The minus sign is intended to give priority to an attractive force.

If the motion is expressed in plane polar coordinates r and θ, that vary as functions of time t, then the acceleration vector $\mathbf{a}$, in terms of the orthonormal basis $(\mathbf{e}_r, \mathbf{e}_\theta)$, is given by

$$\mathbf{a} = (r'' - r\theta'^2)\,\mathbf{e}_r + \frac{1}{r}(r^2\theta')'\,\mathbf{e}_\theta.$$

The dashes denote derivatives with respect to t. Therefore, Newton's second law gives rise to the equations of motion

$$r'' - r\theta'^2 = -F(r), \qquad \frac{1}{r}(r^2\theta')' = 0.$$

We deduce from the second equation that the quantity $r^2\theta'$, physically the angular momentum per unit mass of the orbiting body, is constant. The quantity $r^2\theta'$ has also a geometrical interpretation: it is twice the areal velocity, the latter being the rate at which the radius sweeps out area along the orbit. We have here Kepler's second law, which is therefore valid for motion in any central force field.

We come to the first exercise:

1. For a motion with $r^2\theta' = h$, show that r satisfies the differential equation:

$$\frac{d^2r}{dt^2} = \frac{h^2}{r^3} - F(r) \tag{2.12}$$

Equation (2.12) is a case of Newton's equation. It determines r as a function of time for a motion with $r^2\theta' = h$, given initial conditions specifying r and dr/dt at an initial time. Given a solution $r = R(t)$ the dependence of θ on t is found by recalling that $r^2\theta' = h$, so that we only have to solve the simple differential equation

$$\frac{d\theta}{dt} = \frac{h}{R(t)^2}.$$

It appears from this that the problem of motion in a central force field is solvable by quadratures.

What about the equation of the orbit? We need to express the radius r as a function of the angle θ. This is possible for all motions that are not restricted to the radius, and is the aim of the next exercise.

2. Let $u = 1/r$ and express the motion using θ as independent variable. Show that u satisfies the equation, again an example of Newton's equation:

$$\frac{d^2u}{d\theta^2} = -u + \frac{F(1/u)}{h^2u^2}. \tag{2.13}$$

Hint Obtaining (2.13) from (2.12) and the relation $d\theta/dt = h/r^2$ is an application of the chain rule. It is simplest to carry out the calculations involved (here, and in other cases where a differential equation is transformed by the introduction of new variables) by using Leibniz's notation for derivatives.

Equation (2.13), with initial conditions specifying u and $du/d\theta$ at some initial $\theta = \theta_0$, determines the equation of the orbit in polar coordinates. From any given solution $u = U(\theta)$, and recalling that $\theta' = h/r^2 = hu^2$, we can determine the dependence of the coordinates on time by solving the autonomous separable equation

$$\frac{d\theta}{dt} = h\,U(\theta)^2.$$

A periodic solution of (2.13) does not in general produce a closed orbit in the plane. For this to happen it is necessary and sufficient that the period α of r (or equivalently that of u) as a function of θ is commensurable with π; in other words that α/π is rational. The proof of this is left to the reader. Only in the physically important cases when $F(r) = k/r^2$ (Newtonian inverse square law) and $F(r) = kr$ (restoring force obeying Hooke's law[4]) are all bounded orbits periodic in time.

3. Suppose that along the orbit r is a periodic function, $r = R(\theta)$, of the angle θ, with period α, and that there are integers m and n such that $\alpha m = 2\pi n$. Assume that θ satisfies $d\theta/dt = h/r^2$. Show that the orbit is a closed curve in the plane and the motion along it is periodic in time, with period T given by

$$T = \frac{1}{h}\int_0^{m\alpha} R(\theta)^2\,d\theta.$$

In order to identify the orbit geometrically, some knowledge of coordinate geometry is needed, in particular of conic sections, and their representation in polar coordinates. Using the focus-directrix definition of conic sections, it is easy to show that the equation in polar coordinates of a conic section with eccentricity ϵ and semi-latus-rectum ℓ, is

$$r = \frac{\ell}{1 - \epsilon\cos\theta}$$

given that the origin is one focus and the axis (the line through the focus perpendicular to the directrix) is the x-axis. The semi-latus-rectum is half the chord through the focus normal to the axis, or in our setup the r coordinate at the crossing of the y-axis. The derivation of this formula and the precise setup envisaged are exhibited in Fig. 2.9.

In the case that $0 < \epsilon < 1$ we have an ellipse and the equation can be written

$$r = \frac{a(1-\epsilon^2)}{1 - \epsilon\cos\theta}$$

where a is the semi-major-axis, a quantity of some interest in the case of planetary motion, as it indicates the size of the orbit. The concluding step, expressing d in terms of ϵ and a, is left to the reader. You only have to note that $2a$ is the distance between the two crossings of the x-axis.

In the case of a hyperbola, that is, when $\epsilon > 1$, the two branches can be represented by the same formula if r is allowed to become negative. A purist approach, insisting that r should be positive only, would mean that the formula represents one branch only (the one to the right of the directrix) and a different, similar, formula would be used for the second branch.

[4] For the case $F(r) = kr$ it is easiest to use cartesian coordinates.

Fig. 2.9 Polar equation of a conic

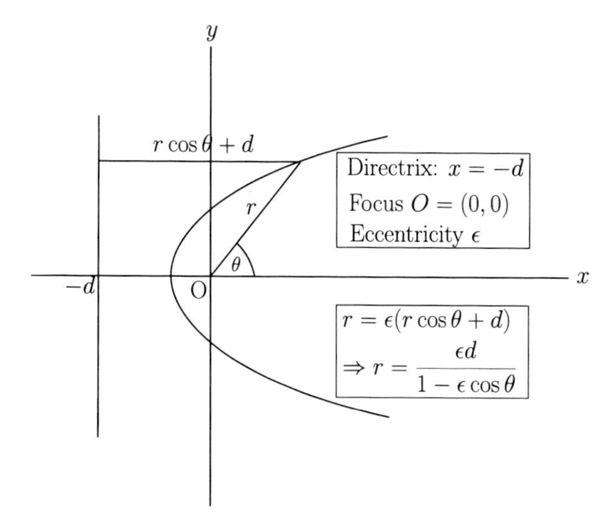

Next we specialise to the case of an inverse square law.

4. Let $F(r) = k/r^2$.

   (a) Show that the general solution of (2.13) is

   $$u = \frac{k}{h^2} - A\cos(\theta + \alpha)$$

   where A and α are arbitrary constants. By rotating the coordinate system (that is, redefining the ray on which θ is zero), we can write the curve as

   $$\frac{1}{r} = \frac{k}{h^2} - B\cos\theta$$

   where $B = |A| \geq 0$. Show that this is a conic section with the origin at one focus and eccentricity $\epsilon = Bh^2/k$. In particular, all bounded orbits are closed curves.

   (b) Show that for each $h \neq 0$ there is a unique circular orbit and find its radius.

   (c) In the case of an elliptical or circular orbit, $0 \leq \epsilon < 1$. Show that the area of the ellipse (or circle) is $\pi h a^{3/2} k^{-1/2}$, where a is the semi-major-axis (radius in the circular case).

   (d) Show that the period of the elliptical orbit is

   $$T = 2\pi a^{3/2} k^{-1/2}.$$

   Hint Use the interpretation of h as twice the areal velocity, and refer to the formula obtained for the area in the previous item. You can also use Exercise 3 but the integral is tricky.

Note The formula shows that the orbital period depends only on the semi-major-axis. The appearance of the $\frac{3}{2}$-power is Kepler's third law. It is remarkable that all bounded orbits with the same semi-major-axis have the same orbital period, even if they have different values of h.

5. A positive number ρ such that $F(\rho) = h^2/\rho^3$ determines a constant solution $u = 1/\rho$ of (2.13). This results in a circular orbit with radius ρ.

   (a) Apply the analysis of case A and case B (Sect. 2.3.2 under the heading 'Critical points') and the terminology used there concerning stable and unstable equilibrium points of Newton's equation, to Eq. (2.13). Show that the constant solution $u = 1/\rho$ (or circular orbit with radius ρ) is stable if

   $$\rho^4 F'(\rho) + 3h^2 > 0$$

   and unstable if

   $$\rho^4 F'(\rho) + 3h^2 < 0$$

   (b) Show that the stability criterion of the previous item is satisfied for the power law $F(r) = r^\gamma$ if $\gamma > -3$ and the instability criterion is satisfied if $\gamma < -3$.

This notion of stability must be interpreted with some care. In the case of a circular orbit with radius ρ, arising from a stable equilibrium point of (2.13), for a given starting point near to it, but not on it, the radial coordinate r oscillates with small amplitude about ρ as a periodic function of θ. This does not necessarily imply that the perturbed orbit is a closed curve (although it is in the case of an inverse square law, as we have seen). However, the perturbed orbit remains close to the initial orbit, exhibiting what is called *orbital stability*. Nevertheless, the motions along the two orbits do not necessarily remain close together, for they will in general have different temporal periods. They are like runners on adjacent tracks who have different speeds and draw apart from one another.

This is illustrated in Fig. 2.10 for the inverse square law. The unperturbed orbit is the circle with centre at the origin, the initial values being $u = 1/\rho$, $du/d\theta = 0$ (the unique ρ for a circular orbit was asked for in Exercise 4(b)). On changing both these values slightly the orbit becomes an ellipse with focus at the origin, but it lies close to the initial circle. Its eccentricity is small (since the unperturbed orbit has zero eccentricity) and to a first approximation it can be treated as a circle whose centre is not exactly at the origin. However, satellites that travel along these two orbits have slightly different periods; so that if they begin side-by-side, they will slowly draw away from each other.

Fig. 2.10 Orbital stability

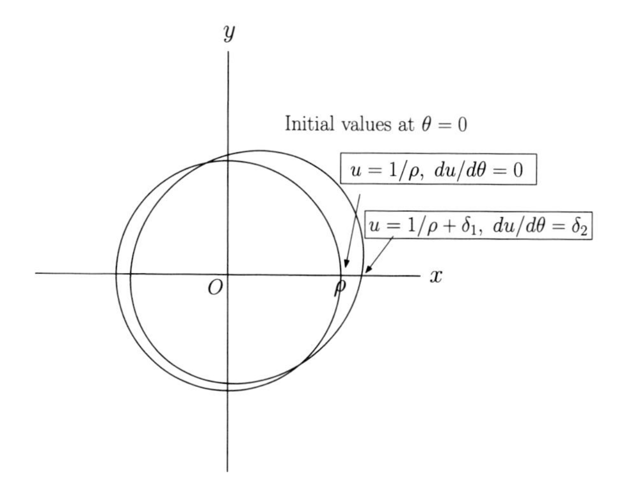

A shortcoming of the discussion of the previous paragraph is that the stability discussed there only compares orbits with the same value of h. What if we slightly perturb the value of h?

6. Let h_0 be an unperturbed value of h, and let ρ_0 satisfy

$$F(\rho_0) = h_0^2/\rho_0^3$$

so that $u = 1/\rho_0$ defines a circular orbit. Assume that the stability criterion

$$\rho_0^4 F'(\rho_0) + 3h_0^2 > 0$$

holds. Show that for each $\varepsilon > 0$ there exists $\delta > 0$, such that for all h that satisfy $|h - h_0| < \delta$, there exists a unique ρ (depending on h) that satisfies $|\rho - \rho_0| < \varepsilon$, and is such that $F(\rho) = h^2/\rho^3$ and $\rho^4 F'(\rho) + 3h^2 > 0$.

The constant solution $u = 1/\rho$ of (2.13), produced by the previous exercise for a perturbed value h, is stable. Therefore, a solution of (2.13) with initial values $du/d\theta = v_0$, $u = u_0$, with h near to h_0, u_0 near to $1/\rho_0$ and v_0 near to 0, will remain close to $1/\rho_0$, because u will remain close to $1/\rho$, which in turn is close to $1/\rho_0$. Thus the orbital stability of the circular orbit $u = 1/\rho_0$ extends to small variations of h as well as variations of the initial values of u and $du/d\theta$.

Chapter 3
Series Solutions of Linear Equations

> What is that word known to all men?

The only large classes of nth order linear equations for which we have a solution method are the equations with constant coefficients and the Euler equations (and these can be transformed into equations with constant coefficients). In this chapter we develop methods based on power series. These methods, apart from their practical value, are a source of new and important functions, for example Bessel functions.

3.1 Solutions at an Ordinary Point

3.1.1 Preliminaries on Power Series

A power series is a function series of the form

$$\sum_{k=0}^{\infty} a_k (x - x_0)^k$$

The constants a_k are called the coefficients (usually real numbers but complex coefficients can be used) and x_0 is called the centre of the series. The term $k = 0$ of the series is always interpreted as the constant function a_0, although the power $(x - x_0)^0$ is not strictly speaking defined when $x = x_0$.

R. Magnus, *Essential Ordinary Differential Equations*, Springer Undergraduate Mathematics Series, https://doi.org/10.1007/978-3-031-11531-8_3

There is a number R, the *radius of convergence*, in the range $0 \leq R \leq \infty$ (we admit the possibility of $R = \infty$ although we don't usually talk of ∞ as a number). We set out the properties associated with the radius of convergence in the following list:

1. The series is absolutely convergent for all x belonging to the open interval $]x_0 - R, x_0 + R[$; or in the case $R = \infty$ for all x.
2. The series is divergent for all x that satisfy $|x - x_0| > R$ (no such points exist if $R = \infty$).
3. Given r, such that $0 < r < R$, the series is uniformly convergent on the closed interval $[x_0 - r, x_0 + r]$.
4. Nothing can be said *a priori* about points satisfying $|x - x_0| = R$ (no such points exist if $R = \infty$).
5. The sum of the series for $x_0 - R < x < x_0 + R$ is a C^∞ function $f(x)$ (which may be extendable outside the interval with nice properties).
6. The successive derivatives of $f(x)$ can be found by differentiating the power series formally term by term. The resulting series are power series with the same radius of convergence.
7. If $R \neq 0$ then the series is the Taylor series of the function it represents, that is, $a_k = f^{(k)}(x_0)/k!$ for $k = 0, 1, 2, \ldots$.
8. The number R can be computed, in principle, by the formula

$$R = \frac{1}{\limsup_{k\to\infty} |a_k|^{\frac{1}{k}}}$$

   The limit is usually tough to evaluate. A more convenient formula in practice is

$$R = \lim_{k\to\infty} \left| \frac{a_k}{a_{k+1}} \right|$$

   when the limit exists or is $+\infty$.
9. A power series with radius of convergence R is convergent for all complex z in the disc $\{z \in \mathbb{C} : |z - x_0| < R\}$ and defines a complex analytic function of z.

Definition A function $f(x)$ of the real variable x defined in an open interval I is called *real analytic* in I if, for each x_0 in I, there is an interval $]x_0 - h, x_0 + h[$ where it is expressible by a convergent power series $\sum_{k=0}^{\infty} a_k(x - x_0)^k$.
A real analytic function is therefore the restriction to the real axis of a complex analytic function. A real analytic function may take complex values; for example e^{ix} is a real analytic function of $x \in \mathbb{R}$.

From results proved in courses on complex variables we obtain some useful properties of real analytic functions:

1. The sum and product of functions real analytic in I are real analytic in I.
2. Let $f(x)$ be real analytic and non-zero everywhere in the interval I. Then the function $1/f(x)$ is real analytic in I.

For example $f(x) := 1 + x^2$ is real analytic in $]-\infty, \infty[$, as a polynomial is obviously a power series with radius of convergence ∞. Since $1 + x^2 \neq 0$ for all real x we have that

$$g(x) := \frac{1}{1+x^2}$$

is also real analytic in $]-\infty, \infty[$. However, for each x_0 the appropriate expansion

$$\frac{1}{1+x^2} = \sum_{k=0}^{\infty} a_k (x - x_0)^k$$

has only finite radius of convergence (it is 1 when $x_0 = 0$).

Exercise How does the radius of convergence depend on x_0?

Answer It is $\sqrt{1 + x_0^2}$.

3.1.2 *Solution in Power Series at an Ordinary Point*

We consider linear equations presented in the form

$$p_2(x)y'' + p_1(x)y' + p_0(x)y = 0, \tag{3.1}$$

where the coefficients $p_j(x)$, $j = 0, 1, 2$, are *real analytic* in an open interval I.

Definition A point $x_0 \in I$ such that $p_2(x_0) \neq 0$ is said to be an *ordinary point* of (3.1). A point $x_0 \in I$ such that $p_2(x_0) = 0$ is said to be an *singular point* of (3.1).

If x_0 is an ordinary point then $p_2(x) \neq 0$ for x in some interval $]x_0 - r, x_0 + r[$, and in this interval we can divide by $p_2(x)$ and reduce (3.1) to standard form

$$y'' + q_1(x)y' + q_0(x)y = 0 \tag{3.2}$$

where $q_1(x)$ and $q_0(x)$ are real analytic in $]x_0 - r, x_0 + r[$. By diminishing r if necessary we may assume that the coefficient functions may be written as power series

$$q_1(x) = \sum_{k=0}^{\infty} a_k (x - x_0)^k, \quad q_0(x) = \sum_{k=0}^{\infty} b_k (x - x_0)^k,$$

both convergent for $x_0 - r < x < x_0 + r$. This means for example that r could be the lesser of the radii of convergence of the two series.

Our object is to show that every solution of (3.2), (and therefore also every solution of (3.1)), on the interval $]x_0 - r, x_0 + r[$, can be written as a power series

$$y(x) = \sum_{k=0}^{\infty} c_k (x - x_0)^k \tag{3.3}$$

convergent in $]x_0 - r, x_0 + r[$, and that the coefficients c_k can be found by solving recurrence equations.

Similar results to the ones we shall study in this chapter hold for the general nth order linear equation with real analytic coefficients. However we shall content ourselves with examining the second order case (the first order case being solvable by quadratures).

Let us first try to find the coefficients in (3.3), by making the assumptions that the solution is given by (3.3) and that the series is convergent. Then we may differentiate (3.3) and substitute into (3.2) thus:

$$\begin{aligned}\sum_{k=0}^{\infty} k(k-1)c_k(x-x_0)^{k-2} \\ + \left(\sum_{k=0}^{\infty} a_k(x-x_0)^k\right)\left(\sum_{k=0}^{\infty} kc_k(x-x_0)^{k-1}\right) \\ + \left(\sum_{k=0}^{\infty} b_k(x-x_0)^k\right)\left(\sum_{k=0}^{\infty} c_k(x-x_0)^k\right) = 0.\end{aligned}$$

Writing out the products in the second and third summands as the Cauchy product of power series, and shifting the summation variable k in the first series, we obtain:

$$\begin{aligned}\sum_{k=0}^{\infty} (k+2)(k+1)c_{k+2}(x-x_0)^k \\ + \sum_{k=0}^{\infty}\left(\sum_{j=0}^{k} (k-j+1)a_j c_{k-j+1}\right)(x-x_0)^k \\ + \sum_{k=0}^{\infty}\left(\sum_{j=0}^{k} b_j c_{k-j}\right)(x-x_0)^k = 0.\end{aligned}$$

This implies the infinite set of equations for the unknown coefficients c_j:

$$(k+2)(k+1)c_{k+2} + \sum_{j=0}^{k}(k-j+1)a_j c_{k-j+1} + \sum_{j=0}^{k} b_j c_{k-j} = 0,$$

$$(k = 0, 1, 2, ...) \qquad (3.4)$$

We see that in the two sums only those c_j with $j \leq k+1$ occur. The equation with $k = 0$ is

$$2c_2 + a_0 c_1 + b_0 c_0 = 0$$

Therefore once we know c_0 and c_1 the equations determine c_j for $j \geq 2$ uniquely and recursively.

What about c_0 and c_1? In fact they can be chosen arbitrarily. This is not surprising for they are the Cauchy data $y(x_0)$ and $y'(x_0)$.

If we can give a proof that the series $\sum_{k=0}^{\infty} c_k(x-x_0)^k$ converges for $x_0 - r < x < x_0 + r$ then we will have proved the following:

Proposition 3.1 *Let the coefficient functions $q_1(x)$ and $q_0(x)$ be given by power series convergent in the interval $x_0 - r < x < x_0 + r$, with coefficients a_k and b_k, and let c_k satisfy the relations* (3.4). *Then the series*

$$\sum_{k=0}^{\infty} c_k(x-x_0)^k$$

is convergent for $x_0 - r < x < x_0 + r$ and its sum is a solution of (3.2).

Note that *some* solutions may have power series that converge on a larger interval. There may, for example, even be some polynomial solutions.

Before we prove the proposition let us give an example showing how to set out the calculations in practice.

Example Find the general solution of

$$y'' - xy = 0.$$

According to Proposition 3.1 every solution is expressible as a power series of the form $y = \sum_{k=0}^{\infty} c_k x^k$, which is convergent for all real x, since here $r = \infty$. Such a series is therefore convergent for all complex x also, and defines an entire analytic function.

Now to the calculation of c_k. We have $a_k = 0$ for all k, $b_k = 0$ for all k except $k = 1$, and $b_1 = -1$. So we have the recurrence equations

$$(k+2)(k+1)c_{k+2} - c_{k-1} = 0.$$

For the purposes of solving it is more convenient to express c_k as a linear combination of c_j with $j < k$, and so we write this as

$$k(k-1)c_k = c_{k-3}. \tag{3.5}$$

It is a useful trick to think of c_k as 0 if k is negative. So the cases $k = 0, 1, 2$ of (3.5) are

$$0.c_0 = 0, \quad 0.c_1 = 0, \quad 2c_2 = 0.$$

These convey the important information that c_0 and c_1 are arbitrary, whilst $c_2 = 0$. Now the relations (3.5) imply that $c_k = 0$ whenever $k \equiv 2 \pmod 3$.

As expected we obtain two independent solutions, $u_1(x)$ and $u_2(x)$, by first choosing $c_0 = 1$, $c_1 = 0$ and then choosing $c_0 = 0$, $c_1 = 1$.

For $u_1(x)$ we have $c_k = 0$ for $k \equiv 1, 2 \pmod 3$. So we only need to consider k as a multiple of 3. Put $k = 3m$. Then we have

$$3m(3m-1)c_{3m} = c_{3m-3} = c_{3(m-1)}$$

that is

$$c_{3m} = \frac{1}{3m(3m-1)} c_{3(m-1)}$$

so we can write down the solution as a product with m factors in front of c_0:

$$\begin{aligned} c_{3m} &= \frac{1}{3m(3m-1)} \frac{1}{(3m-3)(3m-4)} \cdots \frac{1}{3\,.\,2} c_0 \\ &= \frac{1}{3^m\, m!\, (3m-1)(3m-4)\, \ldots\, 2} \end{aligned}$$

We should interpret this as 1 when $m = 0$ (as is usual for the empty product). Avoiding the problem of the empty product we write

$$u_1(x) = 1 + \sum_{m=1}^{\infty} \frac{1}{3^m\, m!\, (3m-1)(3m-4)\ldots 2} x^{3m}.$$

In a similar way we compute $u_2(x)$. Now we have $c_k = 0$ for $k \equiv 0,\ 2 \pmod 3$. So we write $k = 3m + 1$ and find, after a few lines of calculation:

$$c_{3m+1} = \frac{1}{3^m\, m!\, (3m+1)(3m-2)\, \ldots\, 4}$$

Again avoiding the empty product when $m = 0$ we write the solution as

$$u_2(x) = x + \sum_{m=1}^{\infty} \frac{1}{3^m\ m!\ (3m+1)(3m-2)\ \ldots\ 4} x^{3m+1}$$

Proof of Proposition 3.1 To simplify the notation slightly we assume that $x_0 = 0$. There is no real loss of generality since we could make a change of independent variable $t = x - x_0$. We need the following lemma enabling us to compare two sequences satisfying the recurrence relations (3.4).

Lemma 3.1 *Suppose that the coefficients c_k satisfy* (3.4). *Let A_k and B_k be new coefficients and let C_k satisfy* (3.4) *with A_k and B_k replacing a_k and b_k respectively. Now assume that*

i) $|a_k| \le A_k$, $|b_k| \le B_k$ *for* $k = 0, 1, 2, \ldots$
ii) $|c_0| \le C_0$, $|c_1| \le C_1$.

Then $|c_k| \le C_k$ for all k.

Proof The proof is by induction. The conclusion is true for $k = 0, 1$ by assumption. Suppose that it holds up to a certain subscript $k + 1$. Then by (3.4)

$$\begin{aligned}(k+2)(k+1)|c_{k+2}| &\le \sum_{j=0}^{k}(k-j+1)|a_j||c_{k-j+1}| + \sum_{j=0}^{k}|b_j||c_{k-j}| \\ &\le \sum_{j=0}^{k}(k-j+1)A_jC_{k-j+1} + \sum_{j=0}^{k}B_jC_{k-j} \\ &= (k+2)(k+1)C_{k+2}.\end{aligned}$$

□

We continue the proof of Proposition 3.1. Choose a number ρ such that $0 < \rho < r$. Since the series $\sum a_k\rho^k$ and $\sum b_k\rho^k$ are convergent there exist $M > 0$ and $N > 0$, such that

$$\rho^k|a_k| \le M, \quad \rho^k|b_k| \le (k+1)N, \quad (k = 0, 1, 2, \ldots).$$

Now we take

$$A_k = \frac{M}{\rho^k}, \quad B_k = \frac{(k+1)N}{\rho^k}$$

and let C_k be the solutions of (3.4) with A_k replacing a_k, B_k replacing b_k and with starting values $C_0 = |c_0|$, $C_1 = |c_1|$. By the lemma we have $|c_k| \leq C_k$ for all k. But the coefficients C_k are the coefficients of the Maclaurin series of a solution of

$$y'' + Q_1(x)y' + Q_0(x)y = 0 \tag{3.6}$$

where

$$Q_1(x) = \sum_{k=0}^{\infty} A_k x^k = \sum_{k=0}^{\infty} \frac{Mx^k}{\rho^k} = M\left(1 - \frac{x}{\rho}\right)^{-1}$$

and

$$Q_0(x) = \sum_{k=0}^{\infty} B_k x^k = \sum_{k=0}^{\infty} \frac{(k+1)Nx^k}{\rho^k} = N\left(1 - \frac{x}{\rho}\right)^{-2}.$$

So Eq. (3.6) is simply

$$y'' + M\left(1 - \frac{x}{\rho}\right)^{-1} y' + N\left(1 - \frac{x}{\rho}\right)^{-2} y = 0$$

or, even more simply

$$\left(1 - \frac{x}{\rho}\right)^2 y'' + M\left(1 - \frac{x}{\rho}\right)y' + Ny = 0. \tag{3.7}$$

But this is an Euler equation (see Sect. 1.3) and its general solution is of the form

$$y = D_1\left(1 - \frac{x}{\rho}\right)^{\lambda_1} + D_2\left(1 - \frac{x}{\rho}\right)^{\lambda_2}$$

or

$$y = D_1\left(1 - \frac{x}{\rho}\right)^{\lambda} + D_2\left(1 - \frac{x}{\rho}\right)^{\lambda} \ln\left(1 - \frac{x}{\rho}\right).$$

All the functions here have Maclaurin series that converge for $|x| < \rho$. We see, therefore, that $\sum C_k x^k$ converges for $|x| < \rho$, and so, since $|c_k| \leq C_k$, we conclude by comparison of series that $\sum c_k x^k$ converges for $|x| < \rho$. But now we recall that ρ was chosen arbitrarily in the interval $0 < \rho < r$. We conclude finally that $\sum c_k x^k$ converges for $|x| < r$.

3.1.3 Exercises

1. For each of the following equations, determine two independent solutions in the form of power series, convergent in all of $\mathbb{R}$:

   (a) $y'' + xy' + y = 0$
   (b) $y'' - x^2 y = 0$

2. Determine a positive lower bound, with justification, for the radius of convergence of a power series solution $y(x) = \sum_{k=0}^{\infty} c_k x^k$ of each of the following equations, without calculating the solution:

   (a) $(1 + x^2)y'' + y = 0$
   (b) $y'' + (\sin x)y = 0$

   For each equation, calculate the coefficients up to c_6, given that $c_0 = 1$ and $c_1 = 0$.
   Hint For the first equation you can substitute $y(x) = \sum_{k=0}^{\infty} c_k x^k$ directly without first dividing through by $1 + x^2$.
3. The equation

$$y'' - 2xy' + 2\alpha y = 0$$

   where α is a constant, is called Hermite's equation.

   (a) Calculate a solution basis in the form of power series with centre $x = 0$.
   (b) Show that if α is a positive integer or 0 then the equation has a polynomial solution and that the degree of the polynomial is α. (After a suitable normalisation they are called Hermite polynomials.)
   (c) Show that these are the only cases when a non-identically-zero polynomial solution exists.

3.1.4 Projects

A. *Project on the Legendre equation*
The equation in question is

$$(1 - x^2)y'' - 2xy' + \lambda y = 0$$

where λ is a constant. It is often convenient to write it in Sturm-Liouville form (see Sect. 1.2 project B):

$$\left((1 - x^2)y'\right)' + \lambda y = 0.$$

The reader is very likely to encounter Legendre's equation because of its numerous applications in science and technology. To take one example, a complete description of earth's gravity, that takes into account irregularities in its mass distribution and topography, relies on the *Legendre polynomials* (defined below) and the further class of *associated Legendre functions*. This, and numerous other applications, underline the essential nature of Legendre's equation.

A1. Show that a convergent power series $\sum_{n=0}^{\infty} c_n x^n$ satisfies Legendre's equation if and only if the coefficients satisfy the recurrence relations

$$n(n-1)c_n = \big((n-2)(n-1) - \lambda\big)c_{n-2}$$

A2. Show that the general solution on the interval $]-1, 1[$ may be written

$$y(x) = A\phi(x) + B\psi(x),$$

where ϕ and ψ have power series expansions

$$\phi(x) = \sum_{k=0}^{\infty} a_k x^{2k}, \quad \psi(x) = \sum_{k=0}^{\infty} b_k x^{2k+1}$$

A3. Show that if $\lambda = l(l+1)$, where l is a natural number (includes 0), then $\phi(x)$ is a polynomial of degree l if l is even, and $\psi(x)$ is a polynomial of degree l if l is odd.

We see that when $\lambda = l(l+1)$ for some natural number l then Legendre's equation has a polynomial solution of degree l. When this solution is normalised so that $y(1) = 1$, it is called the *Legendre polynomial* of degree l, denoted by $P_l(x)$.

A4. Show that

$$\int_{-1}^{1} P_{l_1}(x)P_{l_2}(x)\,dx = 0$$

if $l_1 \neq l_2$.
Hint. Write the equation in Sturm-Liouville form and use integration by parts.

A5. Show that if the solution $\phi(x)$ (see Exercise A2) is not a polynomial, then its absolute value tends to infinity at both endpoints of the interval $]-1, 1[$. Prove the same for the solution $\psi(x)$.
Hint You might need a delicate convergence test such as Gauss's test to appraise the power series for $\phi(x)$ and $\psi(x)$ at $x = 1$.

A6. Deduce from the previous item that the only solutions of Legendre's equation that are bounded, but not identically zero in the interval $]-1, 1[$, are the Legendre polynomials (up to scalar multiples, of course).

A7. Let

$$\phi_l(x) = \frac{d^l}{dx^l}(x^2-1)^l$$

where l is a natural number. Show that $\phi_l(x)$ satisfies Legendre's equation with $\lambda = l(l+1)$ and is a polynomial of degree l.
Hint. One way to do this is to let $u(x) = (x^2-1)^l$, show that u satisfies the second order equation

$$(1-x^2)u'' + 2(l-1)xu' + 2lu = 0$$

and differentiate l times with the aid of Leibniz's rule for the multiple derivative of a product.

A8. Show that

$$P_l(x) = \frac{1}{2^l\, l!}\frac{d^l}{dx^l}(x^2-1)^l$$

This is often called Rodrigues's formula.

In this project we have only scratched the surface as regards Legendre's equation. The topic will be taken up again in Sect. 7.2 project C. Readers who wish to learn more should consult books on mathematical physics.

3.2 Solutions at a Regular Singular Point

We return to Eq. (3.1) but now assume that the leading coefficient $p_2(x)$ has a zero at x_0, that is, we assume that x_0 is a singular point of the equation.

On dividing by $p_2(x)$ we obtain Eq. (3.2) but now the coefficients $q_1(x) := p_1(x)/p_2(x)$ and $q_0(x) := p_0(x)/p_2(x)$ may have poles at $x = x_0$. In the exceptional case when there are no poles, that is, when the quotients $p_1(x)/p_2(x)$ and $p_0(x)/p_2(x)$ have removable singularities, then x_0 is really no different from an ordinary point.

If there are genuine poles there is, in general, no sense in looking for a solution on an interval that contains x_0. We should look on an open interval that has x_0 as an end-point, $]x_0 - r, x_0[$ or $]x_0, x_0 + r[$. It turns out that there is a class of problems with singular points for which we can describe the behaviour of solutions as x approaches x_0, and develop them in a certain kind of series.

Definition A singular point x_0 of (3.1) is called a *regular singular point* if $q_1(x) := p_1(x)/p_2(x)$ has at worst a simple pole, and $q_0(x) := p_0(x)/p_2(x)$ at worst a double pole, at x_0.

In the case that x_0 is a regular singular point we can write

$$q_1(x) = \frac{Q_1(x)}{x - x_0}, \quad q_0(x) = \frac{Q_0(x)}{(x - x_0)^2}$$

where $Q_1(x)$ and $Q_0(x)$ are analytic in an interval containing x_0. The equation can then be written in the form

$$(x - x_0)^2 y'' + (x - x_0)Q_1(x)y' + Q_0(x)y = 0. \tag{3.8}$$

This serves as a convenient standard form from which it is often plain by inspection that x_0 is a regular singular point.

3.2.1 *The Method of Frobenius*

In the case that x_0 is a regular singular point there is a method, called the method of Frobenius, for developing the solution in a certain kind of series involving ascending powers of $x - x_0$, but it is not an ordinary power series in that there may be non-integer powers. The motivation is the case when Q_1 and Q_0 are constants. We then have an Euler equation, the solutions of which on $]x_0, \infty[$ are spanned by functions of the form $(x - x_0)^\gamma$, values of the *exponent* γ being determined by the indicial equation

$$\gamma(\gamma - 1) + Q_1\gamma + Q_0 = 0.$$

Recall also that if the indicial equation has equal roots than we must use a second solution of the form $(x - x_0)^\gamma \ln(x - x_0)$.

In the case of (3.8) we look for a solution of the form

$$y(x) = (x - x_0)^\gamma \sum_{k=0}^{\infty} c_k (x - x_0)^k = \sum_{k=0}^{\infty} c_k (x - x_0)^{k+\gamma} \tag{3.9}$$

The substitution can be viewed as a version of variation of constants (or parameters), the constant C in the solution $C(x - x_0)^\gamma$ of Euler's equation being replaced by a power series.

Now the series $\sum_{k=0}^{\infty} c_k (x - x_0)^k$ occurring here will be convergent in some open interval $]x_0 - r, x_0 + r[$, but the power function $(x - x_0)^\gamma$ will generally have to be restricted to the interval $]x_0, x_0 + r[$, since γ need not be an integer (in fact γ could even be a complex number).

When we apply (3.9) as a substitution we regard the coefficients c_k, $k = 0, 1, 2, \dots$ as unknown, as well as the exponent γ. Therefore we can always assume that $c_0 \neq 0$ (if c_0 were 0 we could change γ to make c_0 non-zero). The choice $c_0 = 1$ is then logical but does not always lead to the tidiest formula for the solution.

Now let's assume that

$$Q_1(x) = \sum_{k=0}^{\infty} a_k (x - x_0)^k, \quad Q_0(x) = \sum_{k=0}^{\infty} b_k (x - x_0)^k, \quad \big(|x - x_0| < r\big)$$

and substitute (3.9) into (3.8). We obtain

$$\begin{aligned}(x - x_0)^2 \sum_{k=0}^{\infty} (k+\gamma)(k+\gamma-1) c_k (x - x_0)^{k+\gamma-2} \\ + (x - x_0)\left(\sum_{k=0}^{\infty} a_k (x - x_0)^k\right)\left(\sum_{k=0}^{\infty} (k+\gamma) c_k (x - x_0)^{k+\gamma-1}\right) \\ + \left(\sum_{k=0}^{\infty} b_k (x - x_0)^k\right)\left(\sum_{k=0}^{\infty} c_k (x - x_0)^{k+\gamma}\right) = 0,\end{aligned}$$

or more simply,

$$\begin{aligned}\sum_{k=0}^{\infty} (k+\gamma)(k+\gamma-1) c_k (x - x_0)^{k+\gamma} \\ + \left(\sum_{k=0}^{\infty} a_k (x - x_0)^k\right)\left(\sum_{k=0}^{\infty} (k+\gamma) c_k (x - x_0)^{k+\gamma}\right) \\ + \left(\sum_{k=0}^{\infty} b_k (x - x_0)^k\right)\left(\sum_{k=0}^{\infty} c_k (x - x_0)^{k+\gamma}\right) = 0.\end{aligned}$$

Now we multiply out the products of series and write them as series of powers of the form $(x - x_0)^{k+\gamma}$:

$$\begin{aligned}\sum_{k=0}^{\infty} (k+\gamma)(k+\gamma-1) c_k (x - x_0)^{k+\gamma} \\ + \sum_{k=0}^{\infty}\left(\sum_{j=0}^{k} (k - j + \gamma) a_j c_{k-j}\right)(x - x_0)^{k+\gamma} \\ + \sum_{k=0}^{\infty}\left(\sum_{j=0}^{k} b_j c_{k-j}\right)(x - x_0)^{k+\gamma} = 0.\end{aligned}$$

Equating the coefficient of each power $(x - x_0)^{k+\gamma}$ to 0 we obtain the following infinite set of equations

$$(k+\gamma)(k+\gamma-1)c_k + \sum_{j=0}^{k}(k-j+\gamma)a_j c_{k-j} + \sum_{j=0}^{k} b_j c_{k-j} = 0, \quad (k = 0, 1, 2, \ldots).$$

In each sum the subscript in c_{k-j} runs from k down to 0. We peel off the terms with c_k and write them together thus:

$$\Big((k+\gamma)(k+\gamma-1) + (k+\gamma)a_0 + b_0\Big)c_k = P(k+\gamma)c_k,$$

where

$$P(X) = X(X-1) + a_0 X + b_0.$$

So we get the recursive system

$$P(k+\gamma)c_k + \sum_{j=1}^{k}\big((k-j+\gamma)a_j + b_j\big)c_{k-j} = 0, \quad (k = 0, 1, 2, \ldots). \tag{3.10}$$

Usually it's convenient to reverse the order of summation and write this as

$$P(k+\gamma)c_k + \sum_{j=0}^{k-1}\big((j+\gamma)a_{k-j} + b_{k-j}\big)c_j = 0, \quad (k = 0, 1, 2, \ldots). \tag{3.11}$$

The first equation is the one with $k = 0$. The sum is then empty, that is, it is 0, and we find

$$P(\gamma)c_0 = 0.$$

This is where the assumption that $c_0 \neq 0$ comes into play, for we deduce that $P(\gamma) = 0$, that is, γ satisfies the *indicial equation*

$$\gamma(\gamma - 1) + a_0\gamma + b_0 = 0.$$

There are at most two roots. They are called the *exponents at the singular point* x_0.

Let γ_1 be a root of the indicial equation. For the choice $\gamma = \gamma_1$ we attempt to solve the recurrence equations (3.11) using an arbitrary but non-zero value of c_0. We find

$$P(k+\gamma_1)c_k = -\sum_{j=0}^{k-1}\big[(j+\gamma_1)a_{k-j} + b_{k-j}\big]c_j, \quad (k = 1, 2, \ldots), \tag{3.12}$$

which successfully determines c_1, c_2, and so on, provided we nowhere have to divide by 0. Division by 0 will occur if there is an integer $k \geq 1$ such that $P(k + \gamma_1) = 0$, which is to say that $k + \gamma_1$ is *the other root of the indicial equation*. This problem cannot arise unless the roots of the indicial equation differ by a positive integer, and only then if γ_1 is the lower root.

We now state the main conclusion about the solution near a regular singular point.

Proposition 3.2 *Suppose that $Q_1(x)$ and $Q_0(x)$ are given by power series*

$$Q_1(x) = \sum_{k=0}^{\infty} a_k (x - x_0)^k, \quad Q_0(x) = \sum_{k=0}^{\infty} b_k (x - x_0)^k,$$

both convergent for $|x - x_0| < r$. Let γ_1 be a root of the indicial equation

$$X(X - 1) + a_0 X + b_0 = 0$$

and suppose that no further root is of the form $\gamma_1 + k$ where k is a positive integer. Let $c_0 \neq 0$ and let c_k, for $k = 1, 2, \ldots$, be the uniquely determined solutions of the recurrence equations (3.12)*. Then the series $\sum_{k=0}^{\infty} c_k (x - x_0)^k$ is convergent for $|x - x_0| < r$, and the function*

$$y(x) = (x - x_0)^{\gamma_1} \sum_{k=0}^{\infty} c_k (x - x_0)^k$$

is a solution of (3.8) *on the interval* $]x_0, x_0 + r[$.

In order to prove this theorem we only need to establish the convergence of the series with coefficients c_k, where c_k satisfy the recurrence equations (3.11). The following lemma deals with convergence of the series with the coefficients c_k.

Lemma 3.2 *Suppose that the series $\sum a_k X^k$ and $\sum b_k X^k$ are both convergent for $|X| < r$. Suppose that γ is such that $P(\gamma + k) \neq 0$ for any positive integer k. Let c_k satisfy the recurrence equations* (3.11)*. Then the series $\sum c_k X^k$ is convergent for $|X| < r$.*

Proof We give an outline of the proof leaving the reader to fill in the details. We choose some ρ in the interval $0 < \rho < r$, and $M > 0$, such that $\rho^k |a_k| < M$ and $\rho^k |b_k| < M$ for $k = 0, 1, 2, \ldots$. Let $C_k = \rho^k |c_k|$. Then there exists $L > 0$ such that

$$C_k \leq \frac{L}{k} \sum_{j=0}^{k-1} C_j, \quad (k = 1, 2, 3, \ldots)$$

To be precise we can take

$$L = M \sup_{k=1,2,3,\ldots} \frac{k^2 + (|\gamma_1| + 1)k}{|P(k + \gamma_1)|}$$

which is finite because $P(k + \gamma_1)$ is a second degree polynomial in k non-vanishing at all positive integers. Now let D_k satisfy $D_0 = C_0$ and

$$D_k = \frac{L}{k} \sum_{j=0}^{k-1} D_j, \quad k = 1, 2, 3, \ldots$$

By induction we can show $C_k \leq D_k$ for all k. Moreover we can easily show that $D_k/D_{k-1} \to 1$ so that $\sum D_k X^k$ is convergent for $|X| < 1$. Hence $\sum C_k X^k$, which is to say $\sum \rho^k |c_k| X^k$, is convergent for $|X| < 1$, so that $\sum c_k X^k$ is convergent for $|X| < \rho$. Finally we observe that ρ was chosen arbitrarily in the interval $0 < \rho < r$ so we obtain convergence for $|X| < r$. □

The solution given by Proposition 3.2 is called a *solution with exponent* γ_1. In theory, γ_1 could be any number, real or complex; so the solution cannot usually be extended nicely across the singular point. Only when γ_1 is a positive integer or 0 do we get an analytic extension across x_0, but if $\gamma_1 < 0$ we cannot even get a continuous extension as the solution blows up at x_0.

If $\gamma_1 = \alpha + i\beta$ is complex we obtain a complex solution where for $x_0 < x$ we have

$$\begin{aligned}(x - x_0)^{\gamma_1} &= e^{\gamma_1 \ln(x - x_0)} \\ &= (x - x_0)^{\alpha} \Big(\cos\big(\beta \ln(x - x_0)\big) + i \sin\big(\beta \ln(x - x_0)\big)\Big).\end{aligned}$$

If we want a solution on the left-hand interval $]x_0 - r, x_0[$ we can use

$$y(x) = (x_0 - x)^{\gamma_1} \sum_{k=0}^{\infty} c_k (x - x_0)^k$$

and the same coefficients c_k.

If the roots of the indicial equation do not differ by an integer we get two solutions forming a solution basis, by calculating a solution for each root. The proof of independence is left to the reader.

Proposition 3.3 *Suppose that* $Q_1(x)$ *and* $Q_0(x)$ *are given by power series*

$$Q_1(x) = \sum_{k=0}^{\infty} a_k (x - x_0)^k, \quad Q_0(x) = \sum_{k=0}^{\infty} b_k (x - x_0)^k,$$

both convergent for $|x - x_0| < r$. *Suppose that the indicial equation has distinct roots* γ_1 *and* γ_2 *and suppose further that* $\gamma_1 - \gamma_2$ *is not an integer (neither a positive integer, nor a negative integer, nor* 0*). Then the equation has a solution* $u_1(x)$ *with exponent* γ_1 *and a solution* $u_2(x)$ *with exponent* γ_2. *These solutions form a solution basis on the interval* $]x_0, x_0 + r[$.

Example Find a solution basis for the equation

$$2xy'' + y' + xy = 0$$

on the interval $]0, \infty[$.

We see that 0 is a regular singular point by writing the equation in the form

$$x^2y'' + \tfrac{1}{2}xy' + \tfrac{1}{2}x^2y = 0,$$

so that $Q_1(x) = \frac{1}{2}$ and $Q_0(x) = \frac{1}{2}x^2$. Therefore $a_0 = \frac{1}{2}$, $b_2 = \frac{1}{2}$, in all other cases a_k and b_k are 0, and $r = \infty$. We have

$$P(X) = X(X-1) + \tfrac{1}{2}X = X^2 - \tfrac{1}{2}X$$

so the exponents are 0 and $\frac{1}{2}$. The relations (3.11) reduce to

$$(k+\gamma)\big(k+\gamma-\tfrac{1}{2}\big)c_k + \tfrac{1}{2}c_{k-2} = 0.$$

Note that we can also obtain this by substituting $y(x) = \sum_{k=0}^{\infty} c_k x^{k+\gamma}$ directly into the equation $2xy'' + y' + xy = 0$.

For $\gamma = 0$ we find

$$k\left(k - \frac{1}{2}\right)c_k + \frac{1}{2}c_{k-2} = 0$$

that is

$$c_k = -\frac{1}{k(2k-1)}c_{k-2}\,.$$

The case $k = 1$ gives $c_1 = 0$ (think of c_k as 0 if k is negative). It follows that c_k is 0 for all odd k. For even k let $k = 2m$. Then

$$c_{2m} = -\frac{1}{2m(4m-1)}c_{2m-2}$$

so that taking $c_0 = 1$ we have an expression with m factors

$$
\begin{aligned}
c_{2m} &= \Big(-\frac{1}{2m(4m-1)}\Big)\Big(-\frac{1}{(2m-2)(4m-5)}\Big)\dots\Big(-\frac{1}{2\,.\,3}\Big) \\
&= (-1)^m \frac{1}{(2m)(2m-2)\dots 2} \cdot \frac{1}{(4m-1)(4m-5)\dots 3} \\
&= (-1)^m \frac{1}{2^m\, m!\,(4m-1)(4m-5)\dots 3}
\end{aligned}
$$

When $m = 0$ this is the empty product, always interpreted as 1. We get the solution with exponent 0

$$
u_1(x) = 1 + \sum_{m=1}^{\infty} \frac{(-1)^m}{2^m\, m!\,(4m-1)(4m-5)\dots 3}\, x^{2m}
$$

which is an entire analytic function, as the series has infinite radius of convergence, a fact that also follows from the fact that r is infinite.

Next we take $\gamma = \frac{1}{2}$ and find

$$
\Big(k + \frac{1}{2}\Big)kc_k + \frac{1}{2}c_{k-2} = 0
$$

so that

$$
c_k = -\frac{1}{(2k+1)k}c_{k-2}\,.
$$

Again we see that $c_k = 0$ when k is odd and putting $k = 2m$ we get

$$
c_{2m} = -\frac{1}{(4m+1)(2m)}c_{2m-2}
$$

so that, taking $c_0 = 1$ we obtain a product with m factors

$$
\begin{aligned}
c_{2m} &= \Big(-\frac{1}{(4m+1)(2m)}\Big)\Big(-\frac{1}{(4m-3)(2m-2)}\Big)\dots\Big(-\frac{1}{5\,.\,2}\Big) \\
&= \frac{(-1)^m}{\big((2m)(2m-2)\dots 2\big)\big((4m+1)(4m-3)\dots 5\big)} \\
&= \frac{(-1)^m}{2^m\, m!\,(4m+1)(4m-3)\dots 5}
\end{aligned}
$$

We obtain a solution with exponent $\frac{1}{2}$ valid for $0 < x < \infty$:

$$u_2(x) = x^{\frac{1}{2}} \left(1 + \sum_{m=1}^{\infty} \frac{(-1)^m}{2^m \; m! \; (4m+1)(4m-3)\dots 5} \, x^{2m} \right)$$

Again the function within the large parentheses is entire analytic.

This example shows one of the advantages of the method of Frobenius: a large radius of convergence is obtained, in this case infinite. By contrast, if we expanded the solution in powers of $x - 1$, as an ordinary analytic function, we could not guarantee a radius of convergence bigger than 1. The method also yields useful information about the behaviour of solutions as x approaches the singular point.

3.2.2 The Second Solution When $\gamma_1 - \gamma_2$ Is an Integer

It is usually hard to calculate a second solution when $\gamma_1 - \gamma_2$ is an integer. However it is important to know something about its structure, in particular its behaviour as $x \to x_0$.

There are two cases here. The first is when the indicial equation has a double root γ. Then there exists a solution $u_1(x)$ with exponent γ, but a second solution $u_2(x)$, independent of $u_1(x)$, must have a different form.

It is slightly different in the case when $\gamma_1 - \gamma_2$ is a non-zero integer, which we may suppose is positive. Then there exists a solution $u_1(x)$ with exponent γ_1, but the existence of a solution $u_2(x)$ with exponent γ_2 is now uncertain, but, as we shall see, it is not ruled out.

We can approach this problem using the device of reduction of order. The object is to find out the form of the solution. Given the solution $u_1(x)$ with exponent γ_1 we can write a formula for a linearly independent second solution $u_2(x)$ as

$$u_2(x) = u_1(x) \int \frac{e^{-P(x)}}{u_1(x)^2} \, dx$$

where $P(x) = \int (x - x_0)^{-1} Q_1(x) \, dx$ (see Sect. 1.2, Exercise 13). Hence

$$P(x) = \int (x - x_0)^{-1} \sum_{k=0}^{\infty} a_k (x - x_0)^k \, dx = a_0 \ln(x - x_0) + R(x)$$

where $R(x)$ is analytic in a neighbourhood of x_0. Hence

$$e^{-P(x)} = (x - x_0)^{-a_0} S(x)$$

where $S(x) = e^{-R(x)}$ is analytic in a neighbourhood of x_0 and $S(x_0) \neq 0$. Also

$$u_1(x)^{-2} = (x - x_0)^{-2\gamma_1} T(x)$$

where $T(x)$ is analytic in a neighbourhood of x_0 and $T(x_0) \neq 0$. Let $V(x) = S(x)T(x)$. Then we have

$$u_2(x) = u_1(x) \int (x - x_0)^{-a_0 - 2\gamma_1} V(x)\, dx.$$

Let

$$V(x) = \sum_{k=0}^{\infty} v_k (x - x_0)^k.$$

Note that $v_0 \neq 0$. Also note that, by the indicial equation, $\gamma_1 + \gamma_2 = 1 - a_0$, so that $-a_0 - 2\gamma_1 = \gamma_2 - \gamma_1 - 1$. we therefore have

$$u_2(x) = u_1(x) \int (x - x_0)^{\gamma_2 - \gamma_1 - 1} \sum_{k=0}^{\infty} v_k (x - x_0)^k\, dx$$

$$= u_1(x) \int \sum_{k=0}^{\infty} v_k (x - x_0)^{k + \gamma_2 - \gamma_1 - 1}\, dx$$

We emphasise that $v_0 \neq 0$. Now we assume that $\gamma_1 - \gamma_2$ is a non-negative integer m. Then

$$u_2(x) = u_1(x) \sum_{\substack{k=0 \\ k \neq m}}^{\infty} \frac{v_k}{k - m} (x - x_0)^{k-m} + v_m u_1(x) \ln(x - x_0).$$

Now we can make some deductions. Recall that $u_1(x)$ is a solution with exponent γ_1.

Firstly suppose that m is a *positive integer*. There are two subcases, differing by the presence or absence of a logarithm:

Subcase A $\quad v_m = 0$.

In this case $u_2(x)$ takes the form

$$u_2(x) = (x - x_0)^{\gamma_2} \sum_{k=0}^{\infty} d_k (x - x_0)^k$$

with $d_0 \neq 0$, that is, we obtain a solution with exponent γ_2. An alternative interpretation of this case is that the recurrence equations (3.11) are consistent for

$\gamma = \gamma_2$ and a non-zero c_0. This means that on reaching the troublesome equation at $k = m$, the solution of which would seem to require division by 0, we find that

$$\sum_{j=0}^{m-1} \big((j + \gamma_2)a_{m-j} + b_{m-j}\big)c_j = 0.$$

In this case, therefore, c_m is arbitrary and the recurrence equations can be solved. An example of this is given in Exercise 2.

Subcase B $\quad v_m \neq 0$.

In this case $u_2(x)$ takes the form

$$u_2(x) = (x - x_0)^{\gamma_2} \sum_{k=0}^{\infty} d_k (x - x_0)^k + v_m u_1(x) \ln(x - x_0)$$

with $d_0 \neq 0$.

Next consider the case when $m = 0$, that is, the indicial equation has a double root γ. Now the second solution takes the form

$$u_2(x) = (x - x_0)^{\gamma} \sum_{k=1}^{\infty} d_k (x - x_0)^k + v_0 u_1(x) \ln(x - x_0)$$

Summary of the Second Solution

We can summarise the results of this section so that the second solution can be found using a convenient *Ansatz*. In both cases $u_1(x)$ is a solution with exponent γ_1, supposed already known.

1. $\gamma_1 - \gamma_2 = m$ *where* m *is a positive integer.*

We can amalgamate subcases A and B by using the expression

$$u_2(x) = (x - x_0)^{\gamma_2} \sum_{k=0}^{\infty} d_k (x - x_0)^k + D u_1(x) \ln(x - x_0)$$

as a substitution for the purposes of finding a second solution. The unknowns are the coefficients d_j and the multiplier D. However, we must take $d_0 \neq 0$; apart from this d_0 is arbitrary. It turns out that d_m is also arbitrary (can you see why?) and could be put equal to 0. If $D \neq 0$ we could normalise the solution so that $D = 1$ but then d_0 becomes determinate.

2. *The indicial equation has a double root γ.*

In this case there must be a logarithm and we can use the expression

$$u_2(x) = (x - x_0)^{\gamma} \sum_{k=1}^{\infty} d_k (x - x_0)^k + u_1(x) \ln(x - x_0)$$

as a substitution. The unknowns are d_j, $j = 1, 2, 3...$ and they are all determinate. In fact we could have included a term with d_0 but this coefficient is arbitrary (can you see why?) and we can take it to be 0.

3.2.3 The Point at Infinity

The method of Frobenius can be applied to the case of an ordinary point $x = x_0$, as an alternative to the power series substitution used in Sect. 3.1. The reader should verify that it leads to the exponents 0 and 1, and independent power series solutions of the form $\sum_{k=0}^{\infty} c_k (x - x_0)^k$ (with $c_0 \neq 0$) and $\sum_{k=0}^{\infty} d_k (x - x_0)^{k+1}$ (with $d_0 \neq 0$). Of course, this is just the solution basis that we found previously. But it suggests that there may be an advantage in amalgamating the notions of ordinary point and regular singular point. We shall simply use the term *regular point*, to apply to either one. A point that is not a regular point we shall call an *irregular singular point*; about these we shall have nothing more to say.

The substitution $x = 1/t$ transforms the equation

$$p_2(x)\frac{d^2y}{dx^2} + p_1(x)\frac{dy}{dx} + p_0(x)y = 0 \tag{3.13}$$

into the equation

$$t^4 p_2\left(\frac{1}{t}\right)\frac{d^2y}{dt^2} + \left(2t^3 p_2\left(\frac{1}{t}\right) - t^2 p_1\left(\frac{1}{t}\right)\right)\frac{dy}{dt} + p_0\left(\frac{1}{t}\right)y = 0. \tag{3.14}$$

Exercise Verify this claim by carrying out the substitution $x = 1/t$.

Definition We say that infinity (or symbolically ∞) is a regular point of (3.13) when 0 is a regular point of (3.14).

Exercise Show that this is equivalent to requiring the functions

$$\frac{p_1(1/t)}{t\, p_2(1/t)} \quad \text{and} \quad \frac{p_0(1/t)}{t^2\, p_2(1/t)}$$

to be analytic at $t = 0$.

Note that strictly speaking these functions are undefined at $t = 0$ and it is more accurate to say that we require the point $t = 0$ to be a *removable singularity* of the functions. It amounts to saying that there exists $r > 0$, such that they are representable for $0 < |t| < r$ by convergent power series with centre 0.

Now we set

$$q_1(x) = \frac{x\, p_1(x)}{p_2(x)} \quad \text{and} \quad q_0(x) = \frac{x^2\, p_0(x)}{p_2(x)}$$

Then (3.13) has a regular point at infinity precisely when the functions $q_1(x)$ and $q_0(x)$ are *analytic at infinity*, meaning that they can be expanded in series of non-negative integral powers of $1/x$ that converge for $|x|$ sufficiently large. Since Eq. (3.13) can be written

$$x^2 \frac{d^2 y}{dx^2} + x q_1(x) \frac{dy}{dx} + q_0(x) y = 0 \tag{3.15}$$

this gives an easy way to recognise that infinity is a regular point, often with minimal calculation.

Suppose that the functions $q_0(x)$ and $q_1(x)$ can be represented by series of non-negative powers of $1/x$, both of which are convergent for $|x| > r$. We can apply the method of Frobenius to (3.15), using the substitution

$$y = x^\gamma \sum_{n=0}^{\infty} c_n x^{-n}$$

on the assumption that $c_0 \neq 0$ and that γ is unknown, along with the coefficients of the series. This will produce an indicial equation for γ and at least one independent solution in the form of a series of descending powers of x, convergent for $|x| > r$.[1]

Example Solve the equation

$$x^3 y'' + y = 0.$$

Writing the equation as

$$x^2 y'' + \frac{1}{x} y = 0$$

[1] Some sources use $x^{-\gamma}$ in the substitution, leading to some discrepancy about the sign of the exponents at infinity.

we see at once, with reference to the standard form (3.15), that infinity is a regular point. Therefore there exists a solution of the form

$$y = x^{\gamma} \sum_{n=0}^{\infty} c_n x^{-n}$$

with $c_0 \neq 0$, and it will be valid for all $x \neq 0$. Making the substitution and collecting like powers of x lead to the recurrence equations

$$(\gamma - n)(\gamma - n - 1)c_n + c_{n-1} = 0,$$

as the reader should verify. The case $n = 0$ gives the indicial equation

$$\gamma(\gamma - 1) = 0$$

(as usual we think of c_n as 0 when n is negative). The roots are $\gamma = 0$ or 1, and we get the recurrence relations

$$n(n + 1)c_n = -c_{n-1} \quad \text{when } \gamma = 0$$

$$n(n - 1)c_n = -c_{n-1} \quad \text{when } \gamma = 1.$$

The recurrence relations in the case $\gamma = 1$ are inconsistent with the requirement that $c_0 \neq 0$ (why?). The case $\gamma = 0$ gives

$$c_n = \frac{(-1)^n}{n!(n + 1)!}$$

and a first solution

$$u_1(x) = \sum_{n=0}^{\infty} \frac{(-1)^n}{n!(n + 1)!} \frac{1}{x^n}$$

valid for $x \neq 0$. As we saw there is no solution with exponent $\gamma = 1$.

Exercise What is the correct substitution in order to obtain a second solution, independent of u_1?

Answer We can use $y = x \sum_{n=0}^{\infty} d_n x^{-n} + Du_1(x) \ln x$ with $d_0 \neq 0$ and D unknown. Just think of how the solution appears as a function of $t := 1/x$ and refer to the closing paragraphs of Sect. 3.2.2.

3.2.4 Exercises

1. Find a solution basis for each of the following equations on the stated interval:

   (a) $2xy'' + y' + y = 0$ on the interval $]0, \infty[$.
   (b) $2x^2y'' - xy' + (1 + x)y = 0$ on the interval $]0, \infty[$.
   (c) $2(x - 1)y'' - xy' + 3y = 0$ on the interval $]1, \infty[$.
   Hint Expand the solution in ascending powers of $x - 1$. It could simplify things to let $t = x - 1$.

2. Calculate a solution basis for the equation

$$x^2(1 - x^2)y'' - 2y = 0$$

   in two different ways:

   (a) For the interval $0 < x < 1$ using ascending powers of x.
   (b) For the interval $1 < x < \infty$ using descending powers of x.

   Hint You should find in both cases that the exponents differ by an integer; however, in each case both exponents give rise to solutions because the recurrence equations are consistent. Read carefully the description of subcase A in 3.2.2.

3. We continue the study of the Legendre equation

$$(1 - x^2)y'' - 2xy' + \lambda y = 0$$

   initiated in 3.1 project A.

   (a) Show that the Legendre equation has regular singular points at $x = -1$ and $x = +1$ and calculate their exponents.
   (b) Show that the Legendre equation has a solution which is analytic at $x = -1$ and another that is analytic at $x = 1$, neither identically zero.
   (c) When λ is of the form $l(l + 1)$, for some natural number l, then the Legendre equation is satisfied by a polynomial $P_l(x)$ (the Legendre polynomial of degree l, defined in 3.1 project A). Show that in this case a second solution $u(x)$, independent of $P_l(x)$, must blow up as $x \to \pm 1$, and that it does so logarithmically. More precisely, show that $u(x)/\ln(x+1)$ and $u(x)/\ln(1-x)$ have finite non-zero limits as $x \to -1+$ and $x \to 1-$ respectively.
   Note Compare this conclusion with that of 3.1 Exercise 3.1.4.

4. Show that the only singular points of the *hypergeometric equation*

$$x(1 - x)y'' + \big(c - (a + b + 1)x\big)y' - aby = 0$$

   are $x = 0$, $x = 1$ and $x = \infty$, and that they are all regular. Find the exponents at each.

5. Use the substitution

$$y = x \sum_{n=0}^{\infty} d_n x^{-n} + D u_1(x) \ln x$$

with $d_0 \neq 0$, to obtain a second solution to the equation

$$x^3 y'' + y = 0$$

independent of the solution $u_1(x)$ worked out in the text. You should find that d_1 is arbitrary and D non-zero. A general formula for the coefficients is a bit of a challenge and you might be content to define them recursively.

3.2.5 Projects

A. *Project on the Bessel equation*
The equation

$$x^2 y'' + x y' + (x^2 - \mu^2) y = 0 \tag{3.16}$$

is called the Bessel equation of order μ. Here, μ is a real parameter, conventionally called the order, though the terminology is confusing as the equation is second order. We assume, as we clearly can without loss of generality, that $\mu \geq 0$. The solutions of (3.16) are called Bessel functions and play an important role, both in analysis, and in applications to science and technology.

A1. (a) Show that 0 is a regular singular point for Bessel's equation and that the exponents at 0 are μ and $-\mu$.
(b) For a solution with exponent μ or $-\mu$ show that the recurrence equations are

$$k(k + 2\mu)c_k + c_{k-2} = 0, \quad k(k - 2\mu)c_k + c_{k-2} = 0$$

respectively.
(c) Show that $c_k = 0$ if k is odd, and obtain, for a solution with exponent μ, the formula

$$c_{2m} = \frac{(-1)^m}{2^{2m}\, m!\, (m + \mu)(m - 1 + \mu) \ldots (1 + \mu)} c_0$$

The conventional choice for c_0 in item (c) is not 1. It is

$$c_0 = \frac{1}{2^\mu \Gamma(\mu + 1)},$$

using here the Gamma function defined by

$$\Gamma(x) = \int_0^\infty e^{-t} t^{x-1}\, dt, \quad (x > 0).$$

The relevant properties of the Gamma function needed here are:

$$\Gamma(x) = (x-1)\Gamma(x-1) \quad \text{for } x > 1.$$

$$\Gamma(n) = (n-1)! \quad \text{if } n \text{ is a positive integer.}$$

The reader may easily prove them or consult a text on analysis.

A2. (a) The *Bessel function of the first kind of order* μ, usually denoted by $J_\mu(x)$, is the solution with exponent μ and the choice of c_0 stated above. Obtain the formula

$$J_\mu(x) = \sum_{m=0}^{\infty} \frac{(-1)^m}{m!\,\Gamma(m + \mu + 1)} \left(\frac{x}{2}\right)^{2m+\mu}, \quad (0 < x < \infty)$$

In particular if μ is a positive integer or 0

$$J_\mu(x) = \sum_{m=0}^{\infty} \frac{(-1)^m}{m!\,(m + \mu)!} \left(\frac{x}{2}\right)^{2m+\mu}, \quad (0 < x < \infty)$$

(b) Show that if μ is not an integer then a solution exists with exponent $-\mu$ and we can take

$$J_{-\mu}(x) = \sum_{m=0}^{\infty} \frac{(-1)^m}{m!\,\Gamma(m - \mu + 1)} \left(\frac{x}{2}\right)^{2m-\mu}$$

as a second independent solution on $]0, \infty[$.

In case μ is of the form $n + \frac{1}{2}$ the exponents differ by an integer but there is no logarithm in the second solution. Can you see why by looking at the recurrence equations? If μ is an integer a logarithm necessarily appears in the second solution. This is called a *Bessel function of the second kind* and is rather complicated. The reader who needs it should consult the literature (or see the list of classic books at the end of this text).

There may possibly be a thousand and one formulas involving Bessel functions. Here are some, connecting Bessel functions of different orders and their derivatives, that are often useful:

A3. Prove the formulas

$$\frac{1}{x}\frac{d}{dx}\Big(x^{\mu}J_{\mu}(x)\Big) = x^{\mu-1}J_{\mu-1}(x)$$

$$\frac{1}{x}\frac{d}{dx}\Big(x^{-\mu}J_{\mu}(x)\Big) = -x^{-\mu-1}J_{\mu+1}(x).$$

A4. Prove another useful formula:

$$\int_0^a xJ_{\mu}(x)^2\,dx = \frac{1}{2}a^2(J'_{\mu}(a))^2 + \frac{1}{2}(a^2-\mu^2)(J_{\mu}(a))^2.$$

Hint Letting $u(x) = J_{\mu}(x)$ observe that $u(x)$ satisfies

$$\frac{d}{dx}(x^2u'^2 + (x^2-\mu^2)u^2) = 2xu^2.$$

Integrate this formula from 0 to a.

All Bessel functions have infinitely many zeros in the range $0 < x < \infty$, a fact of particular importance for the theory of vibrations. The notes emitted by a circular drum are connected with the zeros of the function $J_0(x)$.

A5. (a) Transform Bessel's equation by setting $y = x^{-\frac{1}{2}}u$ and show that it reduces to

$$u'' + \left(1 + \frac{\frac{1}{4}-\mu^2}{x^2}\right)u = 0. \tag{3.17}$$

(b) Show that every Bessel function $y(x)$ has infinitely many zeros in the interval $]0,\infty[$. Arrange the zeros in an ascending sequence

$$a_1 < a_2 < a_3 < \cdots$$

Show that $a_{j+1} - a_j$ tends to π.
Hint Apply the comparison theorem project (1.2 Exercise B1) to the problem (3.17).

(c) Equation (3.17) also indicates that there are constants A, B, C and D such that

$$J_{1/2}(x) = x^{-1/2}(A\cos x + B\sin x)$$
$$J_{-1/2}(x) = x^{-1/2}(C\cos x + D\sin x).$$

Find them.

Some equations, that cannot be solved in terms of elementary functions, can be integrated using Bessel functions. Thus, Bessel's equation provides us with new and useful transcendental functions, contributing to a list of so-called *special functions.*

A6. Transform Bessel's equation by introducing new variables (t, u) instead of (x, y) where

$$u = x^{-\alpha}y, \quad t = \beta x^{\gamma}$$

with constants α, β, γ. The result is the rather ugly equation

$$\gamma^2 t^2 \frac{d^2u}{dt^2} + \gamma(2\alpha + \gamma)t\frac{du}{dt} + \left(\alpha^2 - \mu^2 + (t/\beta)^{2/\gamma}\right)u = 0.$$

Use this to express a solution basis for the following equations (written, for the reader's ease, in the variables u and t) in terms of Bessel functions:

(a) $\dfrac{d^2u}{dt^2} - tu = 0$

(b) $\dfrac{d^2u}{dt^2} + t^2u = 0$

(c) $\dfrac{d^2u}{dt^2} - t^2u = 0$

There are websites that allow one to find the zeros of Bessel functions and their derivatives; e.g. at time of writing (February 2021) https://keisan.casio.com.

Chapter 4
Existence Theory

> Hold to the now, the here, through
> which all future plunges to the past.

Much of the preceding chapters relied heavily on the unproved proposition asserting the existence of a unique solution to the Cauchy problem for the linear equation, see Proposition 1.3. In this chapter we prove quite general existence theorems, which will, among other things, provide the needed proof of Proposition 1.3.

4.1 Existence and Uniqueness of Solutions

We consider the general first order equation, which may be non-linear. This has the form

$$y' = f(x, y) \tag{4.1}$$

where f is a function of the independent variable x and the dependent variable y. We will need to make some precise assumptions about the domain and regularity of the function f. For a start we shall assume that f *is a continuous function defined in an open connected subset* D *of the plane* $\mathbb{R} \times \mathbb{R}$. Further conditions will be imposed on f where necessary later.

We write the plane as $\mathbb{R} \times \mathbb{R}$, rather than $\mathbb{R}^2$, to emphasise the distinct roles played by the coordinates. Commonly the first coordinate is called the independent variable and the second the dependent variable.

By a solution of (4.1) we will mean a differentiable function $\phi(x)$ defined in some open interval I such that $(x, \phi(x))$ is in D and $\phi'(x) = f(x, \phi(x))$ for all $x \in I$.

R. Magnus, *Essential Ordinary Differential Equations*, Springer Undergraduate Mathematics Series, https://doi.org/10.1007/978-3-031-11531-8_4

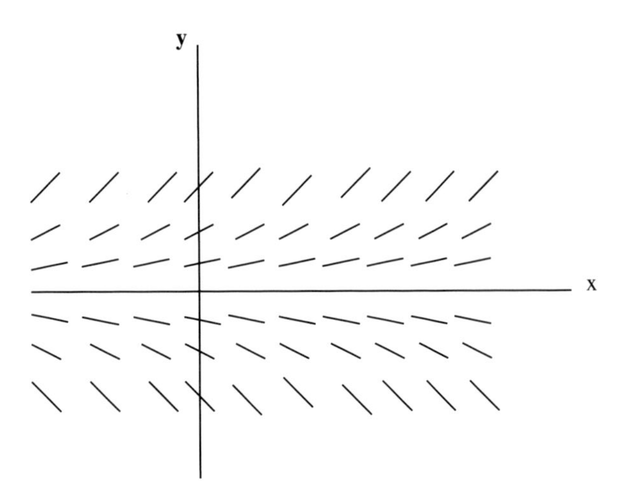

Fig. 4.1 Direction field for $y' = y$

There is a geometrical way of thinking about a solution. Assign to each point (x, y) in D the line with slope equal to $f(x, y)$, or, what amounts to the same thing, the line parallel to the vector $(1, f(x, y))$. It doesn't matter whether you think of the line as passing through (x, y) or not, or whether you think only of a line segment of the right slope or an infinite line. An assignment of a line to each point of D is called a *direction field*[1] in D. This is illustrated in Fig. 4.1.

A solution is a function $\phi(x)$ such that the graph $y = \phi(x)$ has as its tangent line at the point (x, y) the line assigned to (x, y) by the direction field. We refer to the graph $y = \phi(x)$ as a *solution curve* of the differential equation $y' = f(x, y)$.

You will notice that the line with slope $f(x, y)$ cannot be vertical. In more advanced work we might want to allow vertical lines in a direction field. If we admit this possibility, then a plane curve whose tangent line at each point is the line assigned by the direction field to that point, is called an *integral curve* of the direction field. An integral curve is not necessarily the graph of a function in the large. Locally, in the neighbourhood of specific points, it may be expressed by y as a function of x, or, near other points, by x as a function of y. This eliminates the distinction between the independent variable and the dependent variable. A solution curve for the equation $y' = f(x, y)$ is an integral curve for the direction field determined by the equation, but it nowhere has a vertical tangent line. Some of the ideas discussed in this paragraph will reappear in project A, on exact differential equations, and project B, on Lotka-Volterra cycles.

In a general sense, a first order differential equation is the problem of finding integral curves of a given direction field in a domain of $\mathbb{R}^2$. Later we shall see that

[1] This is the conventional name. It might be objected that the line is not assigned a direction; it has a slope only. Therefore 'line field' might be more appropriate. The equivalent notion is important in multidimensional calculus; for example, the assignment of a line in the tangent space at each point of a manifold. This is often called a line field.

an nth order differential equation corresponds to a direction field in $\mathbb{R}^{n+1}$, so it is really the same problem.

4.1.1 Picard's Theorem and Successive Approximations

Suppose that we ask for a solution curve that passes through the given point (x_0, y_0) in D. This is the initial value problem, or Cauchy problem, usually written as a differential equation with initial or Cauchy condition:

$$y' = f(x, y), \quad y(x_0) = y_0 \tag{4.2}$$

A solution to the Cauchy problem will then be a differentiable function $\phi(x)$ defined for x in some open interval I containing x_0, such that for all $x \in I$ the point $(x, \phi(x))$ is in D and satisfies both $\phi'(x) = f(x, \phi(x))$ and $\phi(x_0) = y_0$.

The question arises as to whether the Cauchy problem has a solution, and if it has a solution whether it is unique. The question of uniqueness has to be posed carefully because we might have two solutions defined on different intervals which agree on the intersection of their intervals of definition. In a sense these are not different solutions. We might get around this by extending the solution as far as possible so as to consider only maximally extended solutions that cannot be extended to any larger interval. We would then hope that maximally extended solutions are unique. Alternatively we might simply ask whether the solution is unique on a sufficiently small interval containing x_0. We shall see something of both these approaches.

The main existence theorem for first order differential equations is the following.

Proposition 4.1 (Picard's Existence Theorem) *Let $f : D \to \mathbb{R}$ be continuous in the open, connected set $D \subset \mathbb{R} \times \mathbb{R}$ and let $(x_0, y_0) \in D$. Suppose there exists a rectangle*

$$S = \left\{(x, y) : |x - x_0| \le a, |y - y_0| \le b\right\} \subset D$$

and a constant $K > 0$, such that for each pair of points $(x, y_1), (x, y_2)$ in S the estimate

$$|f(x, y_1) - f(x, y_2)| \le K|y_1 - y_2| \tag{4.3}$$

holds. Let

$$M = \max_{(x,y)\in S} |f(x, y)|$$

and let

$$h = \min\left(a, \frac{b}{M}\right)$$

Then the Cauchy problem (4.2) *has a unique solution defined in the open interval* $I =]x_0 - h, x_0 + h[$.

We preface the proof with some comments about inequality (4.3). The reader may recall from analysis that a function g is called Lipschitz continuous if there is a constant K (a *Lipschitz constant*) such that $|g(t_1) - g(t_2)| \leq K|t_1 - t_2|$ for all t_1 and t_2 in its domain. Therefore inequality (4.3) asserts that the function $f(x, y)$ is Lipschitz continuous in the variable y, uniformly with respect to $(x, y) \in S$. We shall often refer to (4.3) as the *Lipschitz condition* and the constant K as it appears in Proposition 4.1 as the *Lipschitz constant*, though it is not necessarily a Lipschitz constant for f. It is important to appreciate that f, though continuous, need not be a Lipschitz continuous function of (x, y).

The proof of Proposition 4.1 is lengthy and treats existence and unqueness separately.

Proof of Existence The proof of existence proceeds by constructing a function $\phi(x)$, continuous in the closed interval $\bar{I}$, that satisfies the integral equation

$$\phi(x) = y_0 + \int_{x_0}^{x} f(t, \phi(t))\, dt, \quad (x \in \bar{I}) \tag{4.4}$$

By the fundamental theorem of calculus such a function will be differentiable, it will satisfy $\phi'(x) = f(x, \phi(x))$ for each $x \in I$, and the initial condition $\phi(x_0) = y_0$.

We construct a solution of (4.4) by iteration, so called Picard iteration. We define a sequence ϕ_0, ϕ_1, ϕ_2, ... of successive approximate solutions to (4.4) as follows. Firstly

$$\phi_0(x) = y_0, \quad (x \in \bar{I}) \tag{4.5}$$

and inductively

$$\phi_{m+1}(x) = y_0 + \int_{x_0}^{x} f(t, \phi_m(t))\, dt, \quad (x \in \bar{I}) \tag{4.6}$$

In order to define ϕ_{m+1} we have to know that the graph $y = \phi_m(x)$ lies in D for $x \in \bar{I}$. This is guaranteed by the following lemma which shows that the graph is actually trapped in the rectangle S.

Lemma 4.1 *For each m and each $x \in \bar{I}$, $(x, \phi_m(x)) \in S$.*

Proof of the Lemma We use induction. For $m = 0$ and $x \in \bar{I}$ we have $(x, \phi_0(x)) = (x, y_0) \in S$. Suppose the conclusion holds for a given m. Then we may define ϕ_{m+1} and for all $x \in \bar{I}$ we have

$$\phi_{m+1}(x) - y_0 = \int_{x_0}^{x} f(t, \phi_m(t))\, dt.$$

Therefore

$$|\phi_{m+1}(x) - y_0| \leq \left| \int_{x_0}^{x} |f(t, \phi_m(t))|\, dt \right| \leq M|x - x_0| \leq Mh \leq b$$

so that $(x, \phi_{m+1}(x)) \in S$. This proves the lemma. □

Continuation of the Proof of Proposition 4.1 Now we know that each successive approximation has its graph in the rectangle S where the Lipschitz condition (4.3) is valid (see illustration in Fig. 4.2). We next estimate how close ϕ_{m+1} is to ϕ_m.

Lemma 4.2 *For each m and each $x \in \bar{I}$ the estimate*

$$|\phi_{m+1}(x) - \phi_m(x)| \leq \frac{MK^m|x - x_0|^{m+1}}{(m + 1)!} \tag{4.7}$$

holds.

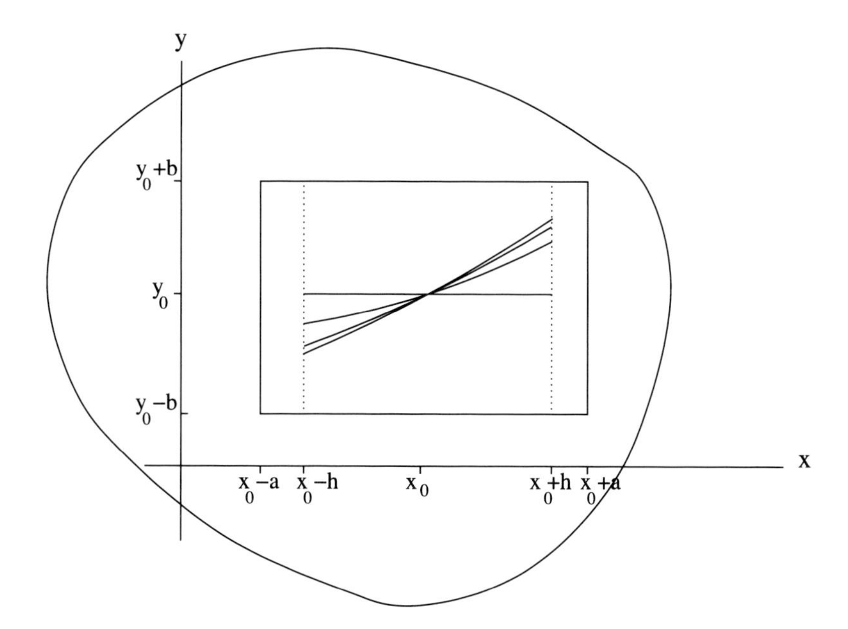

Fig. 4.2 Successive approximations on the interval $\bar{I}$

Proof of the Lemma We use induction. We have

$$\phi_1(x) - \phi_0(x) = \int_{x_0}^{x} f(t, y_0)\, dt$$

so that

$$|\phi_1(x) - \phi_0(x)| \le \left| \int_{x_0}^{x} |f(t, y_0)|\, dt \right| \le M|x - x_0|$$

This is the case $m = 0$ of (4.7). Now we assume that (4.7) holds for a given m. For $x \in \bar{I}$ we have

$$\phi_{m+2}(x) - \phi_{m+1}(x) = \int_{x_0}^{x} \big[f(t, \phi_{m+1}(t)) - f(t, \phi_m(t)) \big]\, dt$$

so that

$$|\phi_{m+2}(x) - \phi_{m+1}(x)| \le \left| \int_{x_0}^{x} \big| f(t, \phi_{m+1}(t)) - f(t, \phi_m(t)) \big|\, dt \right|$$

$$\le K \left| \int_{x_0}^{x} \big|\phi_{m+1}(t) - \phi_m(t)\big|\, dt \right| \le K \left| \int_{x_0}^{x} \frac{MK^m |t - x_0|^{m+1}}{(m+1)!}\, dt \right|$$

$$= \frac{MK^{m+1}|x - x_0|^{m+2}}{(m+2)!}$$

This proves the lemma. □

Continuation of the Proof of Proposition 4.1 From the lemma we find

$$|\phi_{m+1}(x) - \phi_m(x)| \le \frac{MK^m h^{m+1}}{(m+1)!}$$

for all $x \in \bar{I}$. Now the series

$$\sum_{m=0}^{\infty} \frac{MK^m h^{m+1}}{(m+1)!}$$

is convergent (it is an exercise for the reader to find its sum). It follows by the Weierstrass M-test that the series

$$\sum_{m=0}^{\infty} \Big(\phi_{m+1}(x) - \phi_m(x) \Big)$$

is uniformly convergent with respect to $x \in \bar{I}$. Its sum is therefore a continuous function on the interval $\bar{I}$. But we have

$$\sum_{m=0}^{N-1}\Big(\phi_{m+1}(x)-\phi_m(x)\Big)=\phi_N(x)-\phi_0(x)=\phi_N(x)-y_0$$

We conclude that the limit

$$\lim_{N\to\infty}\phi_N(x)$$

exists uniformly for $x \in \bar{I}$ and is a continuous function $\phi(x)$.

We finish the proof of existence by showing that ϕ satisfies (4.4). Now the graph $y = \phi(x)$ must lie in S since $|\phi(x) - y_0| = \lim_{m\to\infty} |\phi_m(x) - y_0| \leq b$. So the Lipschitz condition and the fact that $|\phi(x) - \phi_m(x)|$ converges uniformly to 0 give

$$\begin{aligned}\left|\int_{x_0}^{x} f(t,\phi(t))\,dt-(\phi_{m+1}(x)-y_0)\right| &= \left|\int_{x_0}^{x} f(t,\phi(t))\,dt-\int_{x_0}^{x} f(t,\phi_m(t))\,dt\right|\\ &\leq \left|\int_{x_0}^{x} K|\phi(t)-\phi_m(t)|\,dt\right|\\ &\to 0 \text{ as } m\to\infty\end{aligned}$$

But since $\phi_{m+1}(x)$ converges to $\phi(x)$ we deduce

$$\left|\int_{x_0}^{x} f(t,\phi(t))\,dt-(\phi(x)-y_0)\right|=0$$

that is to say

$$\phi(x)=y_0+\int_{x_0}^{x} f(t,\phi(t))\,dt.$$

This concludes the proof of existence. □

Proof of Uniqueness Suppose there exist two solutions of the Cauchy problem, both on the interval $I =]x_0 - h, x_0 + h[$. Call them ψ_1 and ψ_2.

First we show that the graphs $y = \psi_1(x)$ and $y = \psi_2(x)$ are trapped inside the rectangle S for $x \in I$. Take ψ_1. If the graph could go outside S for some x satisfying $x_0 < x < x_0 + h$ there would be a lowest x in this interval for which $|\psi_1(x) - y_0| = b$. Call this point x_1. In the interval $[x_0, x_1]$ we have $|\psi_1'(x)| = |f(x, \psi_1(x))| \leq M$ and so

$$|\psi_1(x_1)-y_0|=|\psi_1(x_1)-\psi_1(x_0)|\leq M|x_1-x_0|<Mh\leq b.$$

We get a strict inequality $|\psi_1(x_1) - y_0| < b$, contradicting the definition of x_1. So the graph of $y = \psi_1(x)$ cannot exit the rectangle S and the same holds for any solution of the Cauchy problem on the interval I.

Now we have

$$\psi_1(x) - \psi_2(x) = \int_{x_0}^{x} \big[f(t, \psi_1(t)) - f(t, \psi_2(t))\big]\, dt$$

so the Lipschitz condition gives

$$|\psi_1(x) - \psi_2(x)| \le K \left| \int_{x_0}^{x} |\psi_1(t)) - \psi_2(t)|\, dt \right|, \quad (x_0 - h < x < x_0 + h) \tag{4.8}$$

We shall show that this implies $|\psi_1(x) - \psi_2(x)| = 0$ for $x \in I$ by applying the following, widely useful result, usually known as Gronwall's inequality.

Lemma 4.3 *Let ϕ be a continuous function in the interval $[c, d]$, let $k(x)$ be continuous and positive in $[c, d]$ and let m be a constant. Suppose that*

$$\text{(i)} \qquad \phi(x) \le m + \int_{c}^{x} k(t)\phi(t)\, dt, \quad (c \le x \le d).$$

Then

$$\phi(x) \le m \exp\Big(\int_{c}^{x} k(t)\, dt\Big).$$

If, instead of (i), *we assume*

$$\text{(ii)} \qquad \phi(x) \le m + \int_{x}^{d} k(t)\phi(t)\, dt, \quad (c \le x \le d)$$

the conclusion is

$$\phi(x) \le m \exp\Big(\int_{x}^{d} k(t)\, dt\Big).$$

Proof of the Lemma We write the proof for the assumption (i) leaving the other as an exercise. Set

$$U(x) = m + \int_{c}^{x} k(t)\phi(t)\, dt, \quad (c \le x \le d)$$

Then, since $k(x) > 0$, we have

$$U'(x) = k(x)\phi(x) \le k(x)U(x).$$

Hence

$$\frac{d}{dx}\left(e^{-\int_c^x k(t)\,dt}U(x)\right) = e^{-\int_c^x k(t)\,dt}\left(U'(x) - k(x)U(x)\right) \le 0$$

Hence $e^{-\int_c^x k(t)\,dt}U(x)$ decreases in $[c, d]$, giving

$$e^{-\int_c^x k(t)\,dt}U(x) \le U(c) = m,$$

so that

$$\phi(x) \le U(x) \le me^{\int_c^x k(t)\,dt}.$$

This proves case (i) of the lemma. □

Conclusion of the Proof of Theorem 4.1 We apply case (i) of Gronwall's inequality to (4.8) on the interval $[x_0, x_0 + h]$ and case (ii) on the interval $[x_0 - h, x_0]$. In both cases the constant m is 0 so the conclusion is that $|\psi_1(x) - \psi_2(x)| \le 0$, and therefore $\psi_1(x) = \psi_2(x)$, in the interval I. This ends the proof of Theorem 4.1. □

If the reader cares to review the foregoing proof, they should observe that nowhere was it used that M was the *least* upper bound of f in S, only that it was an upper bound. Of course by choosing M as the least upper bound we are making b/M as large as possible and therefore presumably finding the longest interval on which the theorem guarantees a solution. In other cases we might not want to choose the least upper bound, if it then becomes possible to obtain information about a number of different initial values simultaneously.

Consider the case when $|f| \le M$ in some subdomain D_1 of the domain D, and the Lipschitz condition holds in D_1 with constant K, that is,

$$|f(x, y_1) - f(x, y_2)| \le K|y_1 - y_2|, \quad ((x, y_1), (x, y_2) \in D_1).$$

Now choose two numbers a and b, such that $a \le b/M$. For every (x_0, y_0) in D_1, such that the rectangle

$$S = \{(x, y) : |x - x_0| \le a, |y - y_0| \le b\}$$

falls within D_1, we can conclude, by Picard's theorem, that a solution satisfying $y(x_0) = y_0$ exists on the interval $]x_0 - a, x_0 + a[$; the same a for every such (x_0, y_0). Moreover we have the same estimate governing the convergence of the successive approximations on the interval $[x_0 - a, x_0 + a]$, irrespective of the actual initial values, namely:

$$|\phi_{m+1}(x) - \phi_m(x)| \le \frac{MK^m a^{m+1}}{(m+1)!}.$$

The convergence to the solution is therefore uniform with respect to the actual initial conditions, subject to the requirement that the rectangle S falls within D_1; for given ε one can find N, such that beyond the Nth approximation the error is less than ε, irrespective of the initial condition. This observation can be made to yield some important conclusions about the way that the solution depends on the initial condition, a subject that will be taken up later in this text.

One of the things that makes Theorem 4.1 so useful in applications is the fact that the Lipschitz condition (4.3) holds naturally for commonly occurring functions.

Proposition 4.2 *Let $f : D \to \mathbb{R}$ where $D \subset \mathbb{R}^2$ is an open set and suppose that the partial derivative $\frac{\partial f}{\partial y}(x, y)$ exists for all $(x, y) \in D$ and is continuous in D. Then for every closed, bounded and convex set $S \subset D$ there exists K such that* (4.3) *holds for every pair of points (x, y_1) and (x, y_2) in S.*

Proof Let (x, y_1) and (x, y_2) be points in S. Since the line segment L joining them also lies in S the mean value theorem gives

$$|f(x, y_1) - f(x, y_2)| \le \left(\sup_L \left|\frac{\partial f}{\partial y}\right|\right) |y_1 - y_2| \le K|y_1 - y_2|$$

where $K = \sup_S |\partial f/\partial y|$, and K is finite since S is closed and bounded, and $\partial f/\partial y$ continuous. □

It follows from this that if f and $\partial f/\partial y$ are continuous we can apply Proposition 4.1 to any closed rectangle $S = \{(x, y) : |x - x_0| \le a, |y - y_0| \le b\}$ provided it lies within D.

Let's now look at the case of the first order linear equation $y' = p(x)y$. Of course we don't need existence theory since we can solve it by integration, but it is instructive to ask about the Lipschitz condition. Here we have $f(x, y) = p(x)y$ and the domain D is a strip. If $p(x)$ is continuous in the interval $]A, B[$ then

$$D = \{(x, y) : A < x < B\}.$$

Note that A could be $-\infty$ and B could be $+\infty$. The continuous function $p(x)$ need not be bounded in $]A, B[$; it could blow up at an end point. But it is bounded on any closed and bounded subinterval. Let $A < a < b < B$. If $K_{a,b} = \sup_{[a,b]} |p(x)|$ then $K_{a,b} < \infty$ and

$$|f(x, y_1) - f(x, y_2)| \le K_{a,b}|y_1 - y_2|$$

provided only that $a \le x \le b$. Thus f satisfies the Lipschitz condition in the closed substrip

$$S_{a,b} = \{(x, y) : a \le x \le b\}$$

provided $A < a < b < B$.

For equations $y' = f(x, y)$ where f satisfies conditions of this kind we get much stronger results than those of Proposition 4.1.

Proposition 4.3 *Suppose that D is the strip*

$$D = \{(x, y) : A < x < B\}$$

(where A could be $-\infty$ and B could be $+\infty$) and let $f : D \to \mathbb{R}$ be continuous. Suppose that for each a and b such that $A < a < b < B$, there exists $K_{a,b}$ such that the Lipschitz condition

$$|f(x, y_1) - f(x, y_2)| \leq K_{a,b}|y_1 - y_2|$$

holds for every pair of points (x, y_1) and (x, y_2) in the strip

$$S_{a,b} = \{(x, y) : a \leq x \leq b\}.$$

Let $(x_0, y_0) \in D$. Then

1. *The Cauchy problem* (4.2) *has a unique solution on the whole interval* $]A, B[$.
2. *The successive approximations defined by Picard iteration* (4.5) *and* (4.6) *exist on the interval $]A, B[$ and converge everywhere in that interval to the solution.*
3. *On every closed and bounded subinterval $[a, b] \subset\ ]A, B[$ the successive approximations converge uniformly.*

Proof For the definition of successive approximations see Eqs. (4.5) and (4.6). There is no way the graph of $y = \phi_m(x)$ can escape out of D because of the strip geometry. It is therefore obvious that all the successive approximations can be defined in the interval $]A, B[$.

We therefore need to convince ourselves that the result of Lemma 4.2 holds here if we take $I =]a, b[$, where $[a, b] \subset\ ]A, B[$. We only have to redefine M as $\max_{a \leq x \leq b} |f(x, y_0)|$ to start the induction of Lemma 4.2 and the estimate of Lemma 4.2 then follows using the Lipschitz constant $K = K_{a,b}$. The proof of Proposition 4.1 shows that the approximations converge uniformly to a solution on the interval I. The solution obtained on $]a, b[$ is then unique by the same argument as in the proof of Proposition 4.1.

Finally we take sequences $a_j \searrow A$, $b_j \nearrow B$. On each interval $]a_j, b_j[$ we obtain a unique solution, which must extend the solution on the smaller interval $]a_{j-1}, b_{j-1}[$ since the solution on the latter is unique. This collection of solutions gives a solution on $]A, B[$, also unique. □

4.1.2 The nth Order Linear Equation Revisited

Recall the nth order, homogeneous, linear equation

$$y^{(n)} + p_{n-1}(x)y^{(n-1)} + \cdots + p_1(x)y' + p_0(x)y = 0 \tag{4.9}$$

and in particular the Cauchy condition in the form of an *initial vector*, which we write as a column vector:

$$\begin{bmatrix} y(x_0) \\ y'(x_0) \\ \vdots \\ y^{(n-1)}(x_0) \end{bmatrix} = \begin{bmatrix} a_0 \\ a_1 \\ \vdots \\ a_{n-1} \end{bmatrix} \tag{4.10}$$

This suggests that we should look at the *vector function*

$$Y(x) = \begin{bmatrix} y(x) \\ y'(x) \\ \vdots \\ y^{(n-1)}(x) \end{bmatrix}$$

Let's compute its derivative assuming that $y(x)$ is a solution of (4.9), (4.10). We have

$$Y'(x) = \begin{bmatrix} y'(x) \\ y''(x) \\ \vdots \\ y^{(n-1)}(x) \\ y^{(n)}(x) \end{bmatrix} = \begin{bmatrix} y'(x) \\ y''(x) \\ \vdots \\ y^{(n-1)}(x) \\ -p_0(x)y(x) - \cdots - p_{n-1}(x)y^{(n-1)}(x) \end{bmatrix}$$

$$= A(x) \begin{bmatrix} y(x) \\ y'(x) \\ \vdots \\ y^{(n-1)}(x) \end{bmatrix} = A(x)Y(x)$$

where $A(x)$ is the matrix function

$$A(x) = \begin{bmatrix} 0 & 1 & 0 & \dots & 0 \\ 0 & 0 & 1 & \dots & 0 \\ \vdots & \vdots & \vdots & \ddots & \vdots \\ 0 & 0 & 0 & \dots & 1 \\ -p_0(x) & -p_1(x) & -p_2(x) & \dots & -p_{n-1}(x) \end{bmatrix}$$

So the vector function $Y(x)$ satisfies the first order vector differential equation

$$Y'(x) = A(x)Y(x) \tag{4.11}$$

and the Cauchy condition

$$Y(x_0) = \begin{bmatrix} a_0 \\ a_1 \\ \vdots \\ a_{n-1} \end{bmatrix} \tag{4.12}$$

Conversely we shall show that if $Y(x)$ is a solution of (4.11) and (4.12) and if

$$Y(x) = \begin{bmatrix} y_1(x) \\ y_2(x) \\ \vdots \\ y_n(x) \end{bmatrix}$$

then the first coordinate $y_1(x)$ is a solution of (4.9) and (4.10). In fact from the first $n-1$ lines of the matrix $A(x)$ we get

$$y_1'(x) = y_2(x), \quad \dots, \quad y_{n-1}'(x) = y_n(x)$$

giving

$$y_j(x) = y_1^{(j-1)}(x), \quad j = 1, \dots, n$$

and then the last line gives

$$y_n'(x) = -p_0(x)y_1(x) - p_1(x)y_2(x) - \dots - p_{n-1}(x)y_n(x)$$

that is to say

$$y_1^{(n)}(x) = -p_0(x)y_1(x) - p_1(x)y_1'(x) - \dots - p_{n-1}(x)y_1^{(n-1)}(x)$$

so that $y_1(x)$ satisfies (4.9). As for (4.10) we have

$$\begin{bmatrix} y_1(x_0) \\ y_1'(x_0) \\ \vdots \\ y_1^{(n-1)}(x_0) \end{bmatrix} = \begin{bmatrix} y_1(x_0) \\ y_2(x_0) \\ \vdots \\ y_n(x_0) \end{bmatrix} = Y(x_0) = \begin{bmatrix} a_0 \\ a_1 \\ \vdots \\ a_{n-1} \end{bmatrix}$$

This shows that the Cauchy problem (4.9), (4.10) for the nth order equation is precisely equivalent to the Cauchy problem for the first order vector equation (4.11), (4.12).

4.1.3 The First Order Vector Equation

Our task is to prove an existence and uniqueness theorem for the Cauchy problem for the first order vector equation, which we do not assume to be linear. Such an equation has the form

$$y' = f(x, y), \quad y(x_0) = \eta \tag{4.13}$$

where y is vector variable in $\mathbb{R}^n$, f is continuous with values in $\mathbb{R}^n$ and with domain D, an open set in $\mathbb{R} \times \mathbb{R}^n$ (which is the same as $\mathbb{R}^{n+1}$—the notation emphasises the distinct role of the first coordinate). We want to keep the notation as close as possible to that used for the previously treated scalar equation, making changes only where confusion might arise. For example, to avoid confusion we have written the initial vector as η instead of y_0, because of possible, though unlikely, confusion with a coordinate of the vector y. Apart from this the vector problem is almost the same notationally as (4.2); however, the interpretation of the symbols is slightly different.

Of course the vector differential equation in (4.13) is what is often called a system of differential equations; written out in expanded form it is

$$\begin{aligned} y_1' &= f_1(x, y_1, \ldots, y_n) \\ \vdots\ &\quad \vdots \\ y_n' &= f_n(x, y_1, \ldots, y_n) \end{aligned}$$

We can give a geometrical interpretation of the problem $y' = f(x, y)$. To each point $(x, y) = (x, y_1, \ldots, y_n)$ in D we assign the line parallel to the vector $(1, f(x, y)) = (1, f_1(x, y), \ldots, f_n(x, y))$. This produces a direction field in D. A vector valued function $\phi(x)$, defined in an interval I, is a solution of $y' = f(x, y)$ if and only the curve $y = \phi(x)$, in expanded form

$$y_1 = \phi_1(x), \quad \ldots, \quad y_n = \phi_n(x),$$

lies in D, and its tangent line at the point $(x, \phi(x))$ is the line assigned by the direction field at that point.

The fundamental conclusions, Propositions 4.1, 4.2 and 4.3, and their proofs, carry over to the vector case with almost no change of notation. There are slight changes of interpretation. These we now list:

(a) The quantity y is a vector in $\mathbb{R}^n$. To avoid any confusion with coordinates of vectors we use superscripts to distinguish the two vectors in the Lipschitz condition, thus:

$$|f(x, y^{(1)}) - f(x, y^{(2)})| \leq K|y^{(1)} - y^{(2)}|,$$

though there is a slight risk that the superscripts might be confused with derivatives. For the interpretation of the vertical strokes see below.

(b) The domain D is a subset of $\mathbb{R} \times \mathbb{R}^n$.

(c) The function f has values in $\mathbb{R}^n$, and therefore has coordinates. If we wish to indicate them we can write $f = (f_1, ..., f_n)$.

(d) The quantity $|y|$ is the Euclidean length of the vector y:

$$|y| = (|y_1|^2 + \cdots + |y_n|^2)^{1/2}$$

(e) Functions ϕ_m and ϕ occurring in the proof are now vector valued.

(f) The integral of a vector function $\phi = (\phi_1, ..., \phi_n)$ is a vector

$$\int_{x_0}^{x} \phi(t)\,dt = \left(\int_{x_0}^{x} \phi_1(t)\,dt, \ldots, \int_{x_0}^{x} \phi_n(t)\,dt\right).$$

(g) The inequality

$$\left|\int_{x_0}^{x} \phi(t)\,dt\right| \leq \int_{x_0}^{x} |\phi(t)|\,dt, \quad (x_0 < x)$$

is needed for vector functions (the proof is an exercise). The vertical strokes denote the euclidean length.

(h) Weierstrass's theorem on uniform convergence of series is needed for series of vector functions (another exercise).

(i) In Proposition 4.2 we must replace the phrase

the partial derivative $\dfrac{\partial f}{\partial y}(x, y)$ exists for all $(x, y) \in D$ and is continuous in D

by

the partial derivatives $\dfrac{\partial f_j}{\partial y_k}(x, y)$, $j = 1, \ldots, n$, $k = 1, \ldots, n$ exist for all $(x, y) \in D$, and are continuous in D.

(j) The use of the mean value theorem in the proof of Proposition 4.2 needs some rewriting. The following is an exercise. Let $m_k = \sup_S |\nabla f_k(x, y)|$, where $f = (f_1, \ldots, f_n)$ and S is as in Proposition 4.2. Show that

$$|f(x, y^{(1)}) - f(x, y^{(2)})| \leq |m| \, |y^{(1)} - y^{(2)}|$$

where $y^{(1)}$ is a vector $(y_1^{(1)}, \ldots, y_n^{(1)})$, as is also $y^{(2)} = (y_1^{(2)}, \ldots, y_n^{(2)})$, and $m = (m_1, \ldots, m_n)$.
Hint Fix x, $y^{(1)}$ and $y^{(2)}$ and use

$$f_k(x, y^{(1)}) - f_k(x, y^{(2)}) = \int_0^1 \frac{d}{dt} \left(f_k(x, (1-t)y^{(1)} + ty^{(2)}) \right) dt$$

As we have seen the Cauchy problem for the nth order linear equation is equivalent to a Cauchy problem for a first order vector equation

$$y' = A(x)y, \quad y(x_0) = \eta \tag{4.14}$$

where $A(x)$ is an $n \times n$-matrix, whose entries are continuous in the interval I, and η is a given n-vector. To prove the existence of a unique solution on the interval I we must verify the conditions of Proposition 4.3, in its vector version.

So let $[a, b]$ be a bounded subinterval of I. The entries of the matrix $A(x)$ are all bounded on $[a, b]$ since they are continuous functions. Since for any pair of vectors $y^{(1)}$ and $y^{(2)}$ we have $A(x)y^{(1)} - A(x)y^{(2)} = A(x)(y^{(1)} - y^{(2)})$ the Lipschitz condition follows from the following lemma. The proof is an exercise.

Lemma 4.4 *Let $A = (a_{ij})$ be an $n \times n$-matrix. Let m_i be the Euclidean length of the ith row of A. Form the vector $m = (m_1, \ldots, m_n)$. Then for every vector y*

$$|Ay| \leq |m| \, |y|$$

From the lemma and Proposition 4.3 we obtain the following conclusions.

Proposition 4.4 *Suppose that the entries of the matrix $A(x)$ are continuous functions in the open interval I. Let $x_0 \in I$. Then the following conclusions hold:*

1. *The Cauchy problem* (4.14) *has a unique solution on the interval I.*
2. *Picard successive approximations converge to the solution in the interval I.*
3. *In any bounded and closed subinterval $[a, b] \subset I$ the successive approximations converge uniformly.*

It should now be clear that with Proposition 4.4 we have a complete proof of Proposition 1.3.

4.1.4 *Exercises*

1. Consider the Cauchy problem

$$y' = x^2 + y^2, \quad y(0) = 0$$

   (a) Beginning with the approximate solution $\phi_0(x) = 0$ calculate the Picard iterations ϕ_1, ϕ_2 and ϕ_3.
   (b) Choose any convenient rectangle $-a \leq x \leq a$, $-b \leq y \leq b$ and find an explicit value for h such that the Picard iterations are guaranteed by Picard's theorem to converge to a solution on $[-h, h]$.
   (c) What is the highest value for h that can be obtained by varying a and b?

   Note This equation is a special case of the Riccati equation (1.2 Exercise 16).
2. Show that Picard iterations for the problem

$$y' = \sin(xy), \quad y(x_0) = y_0$$

   converge uniformly to a solution on all bounded intervals $[-A, A]$ and that the solution exists on $]-\infty, \infty[$.
3. Prove the following error estimate for the mth Picard approximation, in the notation of Proposition 4.1:

$$|\phi_m(x) - \phi(x)| \leq \frac{K}{M}\frac{(Ka)^{m+1}}{(m+1)!}e^{Ka}, \quad (x_0 - h < x < x_0 + h)$$

4. Consider the equation

$$y' = f(y)$$

   where

$$f(y) = \begin{cases} 2\sqrt{y}, & \text{if } y \geq 0 \\ -2\sqrt{-y}, & \text{if } y < 0. \end{cases}$$

   Show that the following are solution curves:

$$y = (x + C)^2, \quad \text{for } x > -C$$
$$y = -(x + C)^2, \quad \text{for } x > -C.$$

   Deduce that there are infinitely many solutions satisfying the Cauchy-condition $y(0) = 0$ on every interval containing 0. Note that $f(y)$ is continuous, but is not Lipschitz continuous in any neighbourhood of 0.

5. An equation (which could be a vector equation) of the form $y' = f(y)$, where f does not depend on x, is said to be autonomous. Show that if $y(x)$ is a solution of an autonomous equation so are all its translates $y(x + c)$.
   Note In a sense it is enough to develop existence theory for autonomous equations only, since a non-autonomous equation $y' = f(x, y)$ can be thought of as equivalent to the autonomous system

$$\frac{dy}{ds} = f(t, y), \quad \frac{dt}{ds} = 1.$$

6. Consider an equation $y' = f(x, y)$ (which could be a vector equation) where the direction field is periodic with period T. More precisely the domain D of f has the form $\mathbb{R} \times \Omega$, where Ω is open in $\mathbb{R}^n$, and $f(x+T, y) = f(x, y)$ for all $x \in \mathbb{R}$ and $y \in \Omega$. Show that if $y(x)$ is a solution then so is $y(x + T)$.
7. The reformulation of a differential equation as an integral equation can be used to obtain information about the solution of a linear equation as $x \to \infty$. We revisit Bessel's equation, the topic of Sect. 3.2 project A. The first item repeats an exercise from that project. The last gives another approach to proving that Bessel functions have infinitely many zeros. We obtain useful asymptotic information about Bessel functions for large x.

   (a) Let $u(x)$, $x > 0$ satisfy Bessel's equation with index μ (where $\mu \geq 0$). Let $y(x) = x^{\frac{1}{2}} u(x)$. Show that $y(x)$ satisfies

$$y''(x) + \left(1 + \frac{\frac{1}{4} - \mu^2}{x^2}\right) y(x) = 0, \quad (x > 0)$$

   (b) Show that there exist A and B such that

$$y(x) = A \cos x + B \sin x - \int_1^x \left(\frac{\frac{1}{4} - \mu^2}{t^2}\right) y(t) \sin(x - t)\, dt$$

   Hint Use the variations of parameters formula for $y'' + y = g(x)$.
   (c) Show that $y(x)$ is bounded as $x \to \infty$.
   Hint Gronwall's inequality.
   (d) Deduce that there exist C and D, such that

$$\lim_{x \to \infty} \left| y(x) - (C \cos x + D \sin x) \right| = 0.$$

   Hint Show first that the integral in item (b) is convergent at infinity.
   (e) Let $u(x)$ be a solution of Bessel's equation. Show that there exist constants K and α, such that

$$\lim_{x \to \infty} \left| u(x) - K x^{-1/2} \sin(x - \alpha) \right| = 0$$

(f) Deduce that every solution of Bessel's equation tends to 0 as $x \to \infty$, and, if real valued, has infinitely many zeros with the difference between successive zeros tending to π.

4.1.5 *Projects*

A. *Project on exact differential equations*
The differential equation

$$\frac{dy}{dx} = \frac{x+y}{x-y}$$

is defined in the plane with the exclusion of the line $x - y = 0$. However, its solution curves can be extended across this line; they merely have vertical tangents there. It is natural to view the direction field defined by this equation as being defined over the whole plane, except at the origin, and as assigning vertical lines at points on the line $x - y = 0$. This direction field is spanned at the point (x, y) by the vector $(x - y, x + y)$. It is the kernel (or null-space) of the differential 1-form $(x + y)\,dx - (x - y)\,dy$ at the point (x, y). This point of view makes it natural to write the differential equation as

$$(x + y)\,dx - (x - y)\,dy = 0,$$

a form which presents the two variables x and y on equal terms.

In this way, a direction field in a plane domain D, allowing lines that are vertical, may be defined as the kernel of a differential 1-form in D, and presented by writing the equation

$$a(x, y)\,dx + b(x, y)\,dy = 0. \tag{4.15}$$

This 1-form must be non-zero in D, that is, the coefficient functions a and b must not have common zeros. The direction field defined by it is spanned at (x, y) by the non-zero vector $(-b(x, y), a(x, y))$. The integral curves of the direction field, that is, curves that are tangent to the direction field at each point, are the curves on which the 1-form vanishes. They can be built locally by solving the equation

$$\frac{dy}{dx} = \frac{b(x, y)}{a(x, y)}, \qquad y(x_0) = y_0$$

if $a(x_0, y_0) \neq 0$, or else

$$\frac{dx}{dy} = \frac{a(x, y)}{b(x, y)}, \qquad x(y_0) = x_0$$

if $b(x_0, y_0) \neq 0$. Thus through every point there exists a unique locally defined integral curve, provided, for example, the functions $a(x, y)$ and $b(x, y)$ are C^1.

If the differential 1-form $a(x, y)\,dx + b(x, y)\,dy$ is exact, that is to say, if there is a function $F(x, y)$ (a primitive of the 1-form) such that

$$a(x, y)\,dx + b(x, y)\,dy = dF(x, y)$$

then the level curves of F are the integral curves of (4.15). The formula $F(x, y) = C$ can therefore be regarded as a general solution of the problem (4.15).

Recall from multivariate analysis that if a 1-form $a(x, y)\,dx + b(x, y)\,dy$ is exact then it satisfies

$$\frac{\partial b}{\partial x} - \frac{\partial a}{\partial y} = 0,$$

and conversely, that this condition is sufficient for exactness provided the domain is simply connected. A simply connected domain is one in which all closed curves can be continuously deformed to a point. Intuitively, a simply connected domain is without holes. A convenient necessary and sufficient condition for a domain in $\mathbb{R}^2$ to be simply connected is that its complement is connected. All convex sets, and more generally, all star domains, are simply connected.

However, even if the form is not exact, we may be able to find an *integrating factor*. This is a function $m(x, y)$ which is such that the 1-form

$$m(x, y)\big(a(x, y)\,dx + b(x, y)\,dy\big)$$

is exact. Then if $F(x, y)$ is its primitive, we can view the formula $F(x, y) = C$ as a general solution of (4.15). In a simply connected domain, the condition that $m(x, y)$ is an integrating factor is

$$\frac{\partial}{\partial x}(mb) - \frac{\partial}{\partial y}(ma) = 0.$$

After this lengthy preamble we finally come to the first exercises.

A1. (a) Show that x^{-2} is an integrating factor for the 1-form $-y\,dx + x\,dy$ in the half plane $x > 0$. Find the corresponding primitive.

(b) Show that $(x^2 + y^2)^{-1}$ is an integrating factor for the same 1-form, but in the *cut plane*, consisting of $\mathbb{R}^2$ excluding all points of the form $(x, 0)$ with $x \leq 0$. Find the corresponding primitive.

A2. Find an integrating factor for the equation

$$(y^4 - 2y^2)\,dx + (3xy^3 - 4xy + y)\,dy = 0$$

given that there exists one which is a function of xy^2. Find an implicit equation for the integral curves.

In fact, an integrating factor always exists locally, a fact that follows easily from material in Chap. 6, but we do not prove it in this text. The problem is to find one. It was a discovery of S. Lie that if the equation is invariant under a one-parameter group of transformations then an integrating factor can be found by a simple prescription. We refer the reader to Sect. 2.2 for the notion of one-parameter group and invariance of an equation under a group.

Proposition 4.5 *Suppose the direction field defined by the 1-form*

$$a(x, y)\,dx + b(x, y)\,dy$$

is invariant under the group $(\phi_t)_{t\in\mathbb{R}}$ *(written here additively). This is the same as saying that the equation* $dy/dx = -a(x, y)/b(x, y)$ *is invariant under the group. We define the plane vector field*

$$\big(v_1(x, y), v_2(x, y)\big) := \left(\frac{d}{dt}\right)_{t=0} \phi_t(x, y).$$

Then, provided the function $a(x, y)v_1(x, y) + b(x, y)v_2(x, y)$ *is nowhere zero, the function*

$$m(x, y) := \Big(a(x, y)v_1(x, y) + b(x, y)v_2(x, y)\Big)^{-1}$$

is an integrating factor for $a(x, y)\,dx + b(x, y)\,dy$.

If the group is written multiplicatively then the derivative with respect to t is computed at $t = 1$, but the prescription for the integrating factor is otherwise the same.

We will not prove Lie's theorem here, but instead give some examples of integrating factors suggested by it, and invite the reader to verify that they are indeed integrating factors. The results of 2.2 Exercise 4 are relevant.

A3. Let the equation $dy/dx = f(x, y)$ be invariant under the group of dilatations $\phi_t(x, y) = (tx, ty)$, $(t > 0)$. Show that Lie's theorem says that $(xf(x, y) - y)^{-1}$ is an integrating factor for the 1-form $f(x, y)\,dx - dy$ and verify that this is the case. Use this to solve the equation

$$\frac{dy}{dx} = \frac{x + y}{x - y}.$$

A4. Let the equation $dy/dx = f(x, y)$ be invariant under the rotation group. Show that Lie's theorem says that $(yf(x, y) + x)^{-1}$ is an integrating factor for the 1-form $f(x, y)\,dx - dy$ and verify that this is the case. Use this to solve the equation

$$\frac{dy}{dx} = \frac{x + y}{x - y}.$$

A5. (a) The group of weighted dilatations is defined by $\phi_t(x, y) = (tx, t^\lambda y)$, $(t \in \mathbb{R}_+)$. Show that the equation $dy/dx = f(x, y)$ is invariant under the group of weighted dilatations if and only if $f(tx, t^\lambda y) = t^{\lambda-1} f(x, y)$ for all $(x, y) \neq (0, 0)$ and $t > 0$.

(b) Show that the equivalent partial differential equation (compare 2.2 Exercise 4) is

$$x\frac{\partial f}{\partial x} + \lambda y\frac{\partial f}{\partial y} = (\lambda - 1)f, \qquad \big((x, y) \in D\big).$$

(c) Show that if the equation $dy/dx = f(x, y)$ is invariant under the group of weighted dilatations then Lie's theorem indicates that the function $(xf(x, y) - \lambda y)^{-1}$ is an integrating factor for the 1-form $f(x, y)\,dx - dy$. Verify that this is the case.

B. *Project on Lotka-Volterra cycles*

The evolution of a system with two degrees of freedom can be described by the autonomous equation pair

$$\frac{dx}{dt} = f(x, y), \quad \frac{dy}{dt} = g(x, y).$$

Here we have as dependent variable the plane coordinate vector (x, y), whilst the independent variable t would commonly be time. A solution $(x(t), y(t))$ is usually seen as a plane curve parametrised by time, and as such can be called a phase curve of the system. The (x, y)-plane would then be called the phase plane. This is analogous to the usage we encountered in connection with Newton's equation in Chap. 2.

Such models for the time-evolution of two interdependent quantities are very frequently encountered in applications. In this text there is no space to give a general account of these problems. However, we shall take a look at one famous example.

Firstly we note that if we disregard the parametrisation of the phase curve, and consider it to be a curve in $\mathbb{R}^2$, a geometrical object so to speak, then the 1-form

$$\frac{dx}{f(x, y)} - \frac{dy}{g(x, y)}$$

is zero on it. We can use methods from project A, on exact differential equations, to determine the integral curves of this 1-form, which would reveal the phase curves but not their parametrisation by time.

The Lotka-Volterra model for the interaction of two competing species in the wild, whose quantities are given by the coordinates x and y, is the system

$$\frac{dx}{dt} = x(A - By), \quad \frac{dy}{dt} = y(-C + Dx)$$

where A, B, C and D are positive constants. The rationale behind this is that the species whose quantity is x is preyed upon by the species whose quantity is y. This accounts for the terms $-Bxy$ in the first equation and Dxy in the second. The other terms suggest that in the absence of interactions the predator dies out whilst the prey thrives. This, admittedly naive, model has been suggested to describe the interaction of falcons and ptarmigans in Iceland. We shall ignore its shortcomings.

In accordance with the practical interpretation we consider the Lotka-Volterra system in the first quadrant, $x > 0$, $y > 0$.

B1. Show that there is a unique constant solution in the first quadrant, given by $(x, y) = (C/D, A/B)$.

The obvious interpretation of this solution is that the two populations are balanced so that their quantities do not change with time. If we allow x or y to be 0 then there is another constant solution $(x, y) = (0, 0)$, with an obvious interpretation. The main interest lies in the non-constant solutions in the quadrant $x > 0$, $y > 0$.

B2. (a) Show that the 1-form

$$\left(\frac{-C + Dx}{x}\right) dx - \left(\frac{A - By}{y}\right) dy,$$

defined for $x > 0$ and $y > 0$, is zero on the phase curves and is exact.

(b) Show that the function

$$F(x, y) := Dx - C \ln x + By - A \ln y, \quad (x > 0, \ y > 0)$$

is a primitive of the 1-form in item (a).

The level curves of $F(x, y)$ reveal the phase curves, apart from parametrisation by time.

B3. (a) Show that the function $F(x, y)$ is strictly convex. In fact, it is the sum of a strictly convex function of x and a strictly convex function of y.

(b) Show that

$$\lim_{x\to\infty} F(x, y) = \infty, \quad \lim_{x\to 0+} F(x, y) = \infty,$$

$$\lim_{y\to\infty} F(x, y) = \infty, \quad \lim_{y\to 0+} F(x, y) = \infty$$

the first and second limits being attained uniformly with respect to y, and the third and fourth uniformly with respect to x.

(c) Show that $F(x, y)$ attains its minimum at the point $(C/D, A/B)$.

(d) Show that all level curves of $F(x, y)$ at a level above its minimum value are closed curves encircling $(C/D, A/B)$ and having convex interior domain.

Hint Item (d) is arguably tricky to do rigorously. One approach is to show first that a level curve Γ, given say by $F(x, y) = h$, where $h > \min F$, is a differentiable curve. This is just the implicit function theorem. By item (b) it is confined to a square $\varepsilon \le x \le k$, $\varepsilon \le y \le k$ (for some $\varepsilon > 0$ and $k > 0$). For every line ℓ through $(C/D, A/B)$ there is exactly one point of Γ on ℓ on each side of $(C/D, A/B)$. This follows from the strict convexity of $F(x, y)$ and item (b). Finally one can take polar coordinates r and θ with origin $(C/D, A/B)$, and describe Γ by an equation $r = R(\theta)$, where $R(\theta)$ is 2π-periodic and continuous. This exhibits an explicit homeomorphism from Γ to a circle.

B4. Show that the motion along the level curves of the previous item is time-periodic.
Hint The solution $(x(t), y(t))$ has to go all the way round its level curve and return to its starting point after a finite time.

A time-periodic non-constant phase curve is called a *cycle*. The level curves of $F(x, y)$ are the *Lotka-Volterra cycles*. The period depends on the level, but there seems to be no nice formula for it. The interested reader should consult the literature on the subject. Figure 4.3 shows Lotka-Volterra cycles drawn by the graph-plotting application desmos for the case $A = B = C = D = 1$. The reader may figure out the correct direction of motion, which the author lacked the skill to indicate in the diagram.

We conclude this project by presenting a condition that excludes the presence of cycles in an autonomous plane system.

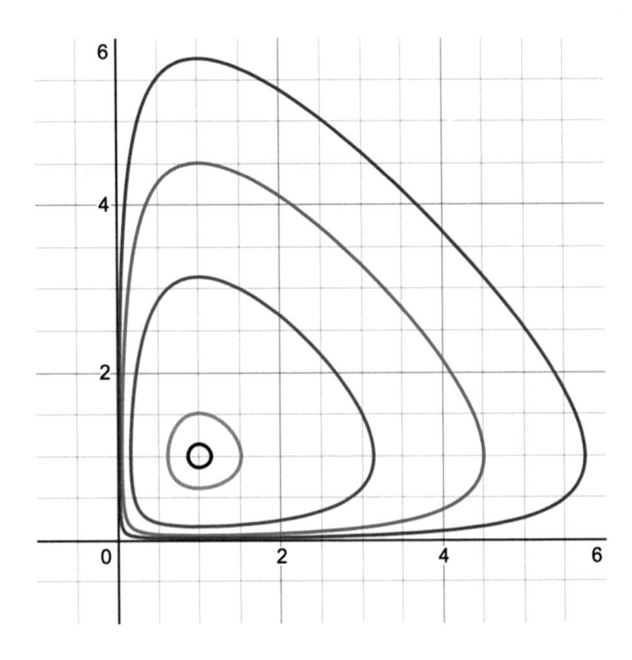

Fig. 4.3 Lotka-Volterra cycles

B5. Prove Bendixson's negative criterion. Consider the plane system

$$\frac{dx}{dt} = f(x, y), \quad \frac{dy}{dt} = g(x, y).$$

where the functions f and g are C^1. Suppose that in a simply connected region A the function

$$\frac{\partial f}{\partial x} + \frac{\partial g}{\partial y}$$

is either everywhere strictly positive or everywhere strictly negative. Then there can be no cycle (non-constant periodic phase curve) in A.
Hint Along the phase curve we have $g\,dx - f\,dy = 0$.
Note The quantity, which is assumed to be entirely of one sign in A, is the divergence of the vector field (f, g). A more general version, the Bendixson-Dulac theorem, assumes that there exists a function ϕ, such that the divergence of the vector field $(\phi f, \phi g)$ is of one sign in A. The reader might like to prove it.

C. *Project on Osgood's uniqueness theorem*
A modulus of continuity is a function $\omega : [0, \infty[\to [0, \infty[$ that is continuous, strictly increasing and satisfies $\omega(0) = 0$. The name arises when it is used to give quantitative bounds for the continuity of a function, by means of the inequality

$$|f(x_1) - f(x_2)| \leq \omega(|x_1 - x_2|).$$

Lipschitz continuity corresponds to the case $\omega(t) = Kt$.

The following proposition and corollary establish uniqueness of the solution of the initial value problem, under conditions weaker than the Lipschitz condition of Proposition 4.1.

Proposition 4.6 *Let D be an open, connected subset of $\mathbb{R} \times \mathbb{R}^n$. Suppose that the function $f : D \to \mathbb{R}^n$ is continuous and satisfies*

$$\left|f(x, y^{(1)}) - f(x, y^{(2)})\right| \leq \omega\left(|y^{(1)} - y^{(2)}|\right),$$

for each pair $(x, y^{(1)})$ and $(x, y^{(2)})$ in D, with modulus of continuity ω. Suppose further that the integral $\int_0^1 1/\omega$ is divergent at 0.

Let $\phi_1(x)$ and $\phi_2(x)$ be solutions of $y' = f(x, y)$ on the same open interval J. If there exists $x_0 \in J$, such that $\phi_1(x_0) \neq \phi_2(x_0)$, then $\phi_1(x) \neq \phi_2(x)$ for every point $x \in J$.

Corollary 4.1 (Osgood's Uniqueness Theorem) *Under the conditions of the previous proposition, if $(x_0, \eta) \in D$ and if a solution of the Cauchy problem $y' = f(x, y)$, $y(x_0) = \eta$ exists,[2] then it is unique.*

C1. Prove Proposition 4.6, and hence Osgood's uniqueness theorem, in a number of steps. We suppose that $x_0 \in J$ and $\phi_1(x_0) \neq \phi_2(x_0)$. We let $\psi(x) = \phi_1(x) - \phi_2(x)$ for each $x \in J$. Then $\psi(x_0) \neq 0$. The object is to show that $\psi(x) \neq 0$ for all $x \in J$.

(a) Show that if $\psi(x) \neq 0$, then

$$\left| \frac{d}{dx} |\psi(x)| \right| \leq \omega(|\psi(x)|).$$

(b) Let $\Omega(s)$ be an antiderivative of $1/2\omega(s)$ on the interval $0 < s < \infty$. Show that $\lim_{s \to 0+} \Omega(s) = -\infty$.

(c) Let $\Gamma_+(x, s) = x + \Omega(s)$, $\Gamma_-(x, s) = x - \Omega(s)$ for $s > 0$ and $x \in J$. As long as $\psi(x) \neq 0$, show that

$$\frac{d}{dx}\Gamma_+(x, |\psi(x)|) > \frac{1}{2}, \qquad \frac{d}{dx}\Gamma_-(x, |\psi(x)|) > \frac{1}{2}$$

(d) Show that the functions $\Gamma_+(x, s)$ and $\Gamma_-(x, s)$ are increasing along the curve $s = |\psi(x)|$ in the (x, s)-plane, in any open interval in which $\psi(x) \neq 0$.

(e) Conclude the proof of Proposition 4.6. Show that $\psi(x)$ cannot reach 0 in the interval J, neither above x_0 nor below it. In other words $\phi_1(x) \neq \phi_2(x)$ for all $x \in J$.

[2] A solution does exist. That is a consequence of Peano's existence theorem, a result that is beyond the scope of this text.

Chapter 5
The Exponential of a Matrix

> I am, a stride at a time. A very short space of time through very short time of space.

This chapter is mostly devoted to studying the homogeneous linear system with constant coefficients, in other words the problem (4.14) for the case when $A(x)$ is a constant matrix A. The initial value problem

$$y' = Ay, \quad y(x_0) = \eta \tag{5.1}$$

is naturally considered on the whole real line $\mathbb{R}$. Later in the chapter we shall look at the general case of linear equations with variable coefficients.

5.1 Defining the Exponential

The constant coefficient problem is solved by the exponential function. The reader should recall the Picard iterations defined in the proof of Proposition 4.1.

Lemma 5.1 *The sequence* $(\phi_m)_{m=0}^{\infty}$ *of Picard approximations to the solution of* (5.1) *is given by*

$$\phi_m(x) = \left(\sum_{k=0}^{m} \frac{(x-x_0)^k}{k!} A^k\right) \eta,$$

where A^0 is interpreted as the $n \times n$ unit matrix I_n.

R. Magnus, *Essential Ordinary Differential Equations*, Springer Undergraduate Mathematics Series, https://doi.org/10.1007/978-3-031-11531-8_5

Proof We use induction. The formula obviously holds for $m = 0$. Suppose it holds for a given m. We derive it for $m + 1$ as follows:

$$\begin{aligned}\phi_{m+1}(x) &= \eta + \int_{x_0}^{x} A\phi_m(t)\,dt \\ &= \eta + \int_{x_0}^{x} A\left(\sum_{k=0}^{m} \frac{(t-x_0)^k}{k!}A^k\right)\eta\,dt \\ &= \eta + A\left(\sum_{k=0}^{m} \frac{(x-x_0)^{k+1}}{(k+1)!}A^k\right)\eta \\ &= \left(\sum_{k=0}^{m+1} \frac{(x-x_0)^k}{k!}A^k\right)\eta\end{aligned}$$

□

In conjunction with (5.1) we can also consider the *matrix differential equation*

$$\frac{dT}{dx} = AT, \quad T(0) = I_n \tag{5.2}$$

where the dependent variable T is an $n \times n$ matrix and I_n is the $n \times n$ unit matrix. We can view this as a linear differential equation in the vector space $\mathbb{R}^{n^2}$ and therefore it has a unique solution defined on all of $\mathbb{R}$.

Definition The exponential $\exp(A)$ of the matrix A is defined to be the value at $x = 1$ of the solution of the initial value problem (5.2). It is less formally, but more commonly, denoted by e^A.

Proposition 5.1 *The solution of* (5.2) *is* e^{xA}.

Proof We temporarily denote the solution of (5.2) by $\Phi(x, A)$. For a given x_0 the chain rule yields

$$\frac{d}{dx}\Phi(x_0x, A) = (x_0A)\Phi(x_0x, A),$$

and since $\Phi(x_0x, A)$ reduces to I_n when $x = 0$ we infer that $\Phi(x_0x, A) = \Phi(x, x_0A)$. Putting $x = 1$ we conclude that $\Phi(x_0, A) = \Phi(1, x_0A) = e^{x_0A}$. □

Using Lemma 5.1, it is easy to see that the mth order Picard approximation to e^{xA} is

$$\sum_{k=0}^{m} \frac{x^k}{k!}A^k$$

Using Proposition 4.4 we now obtain two important conclusions.

Proposition 5.2 *Let A be an $n \times n$-matrix. Then:*

1. *The exponential function satisfies*

$$e^{xA} = \sum_{k=0}^{\infty} \frac{x^k}{k!} A^k,$$

where the matrix series converges for all x, uniformly on every bounded interval.

2. *The solution of the Cauchy problem* (5.1) *is given by*

$$y(x) = e^{(x-x_0)A}\eta.$$

The calculation of e^{xA} is an algebraic problem and is essentially solved once the eigenvalues of A are known.

5.1.1 Exercises

1. Prove the following properties of e^{xA} directly from the fact that e^{xA} is the solution of the initial value problem $T' = AT$, $T(0) = I_n$. Hints are given for the first two; similar methods can be used for the rest.
   (a) $e^{(x_1+x_2)A} = e^{x_1A}e^{x_2A}$.
   Hint Consider the function $x \mapsto e^{(x+x_2)A}$.
   (b) If A and B are matrices and $AB = BA$ then $Ae^B = e^BA$.
   Hint Consider the function $x \mapsto Ae^{xB}$.
   (c) If $AB = BA$ then $e^Ae^B = e^Be^A$.
   (d) If $AB = BA$ then $e^{A+B} = e^Ae^B$.
   (e) If T is an invertible matrix then $T^{-1}e^AT = e^{T^{-1}AT}$.
2. Let D be the diagonal matrix $\text{diag}(d_1, \dots, d_n)$, comprising the numbers d_1, $\dots$, d_n on the diagonal and zeros in all other places. Show that $e^{xD} = \text{diag}(e^{xd_1}, \dots, e^{xd_n})$. Deduce that if a matrix A is diagonalisable, and if $T^{-1}AT = \text{diag}(d_1, \dots, d_n)$ then $e^{xA} = T^{-1}\text{diag}(e^{xd_1}, \dots, e^{xd_n})T$.
3. Calculate e^{xA} in the following cases using the exponential series:

$$A = \begin{bmatrix} 0 & 1 \\ 0 & 0 \end{bmatrix}, \quad \begin{bmatrix} 0 & -1 \\ 1 & 0 \end{bmatrix}, \quad \begin{bmatrix} 0 & 1 & 0 \\ 0 & 0 & 1 \\ 0 & 0 & 0 \end{bmatrix}$$

4. Suppose that the $n \times n$ matrix A satisfies $A^2 = -I_n$. Show that

$$e^{xA} = (\cos x)I_n + (\sin x)A.$$

5.2 Calculation of Matrix Exponentials

We are going to study some commonly used methods for calculating the exponential e^{xA} given that A is an $n \times n$ matrix. All the methods we consider here begin from the eigenvalues of A, including the complex eigenvalues. Calculating the eigenvalues is the difficult part. We only consider here what to do once the eigenvalues are known.

Recall that the matrix function e^{xA} is the solution $T = e^{xA}$ of the initial value matrix problem (5.2), and that the solution of (5.1) is given by $y = e^{(x-x_0)A}\eta$. In the case that A has complex eigenvalues we may have to consider the initial value problem (5.1) with a complex vector η, in which case the solution is still given by the same formula. The extension to complex initial vectors is most easily understood by thinking of $\mathbb{C}^n$ as the complexification of $\mathbb{R}^n$, that is, the sum $\mathbb{R}^n \oplus i\mathbb{R}^n$. For a complex vector $u + iv$ we have $A(u + iv) = Au + iAv$ and $e^{xA}(u + iv) = e^{xA}u + ie^{xA}v$.

We shall describe two principal methods that are used to calculate e^{xA}. The first may be called the eigenvector method. The second is really a group of methods, which we can loosely associate with the Cayley-Hamilton theorem.

5.2.1 *Eigenvector Method*

A basic observation is that if $M(x)$ is a matrix function that satisfies $M'(x) = AM(x)$, and if the matrix $M(0)$ is invertible, then $e^{xA} = M(x)M(0)^{-1}$. We only have to observe that the matrix function $M(x)M(0)^{-1}$ solves the problem (5.2).

The equation $M'(x) = AM(x)$ says nothing more than that each column of the matrix function $M(x)$ is a vector solution of $y' = Ay$. And the condition that $M(0)$ is an invertible matrix simply says that the n solutions formed by the columns are linearly independent. So if we have a basis $\{v_1, \dots, v_n\}$ of $\mathbb{R}^n$ (or of $\mathbb{C}^n$ if there are complex eigenvalues), then the solutions of the n initial value problems

$$\frac{dy}{dx} = Ay, \quad y(0) = v_k, \quad (k = 1, 2, \dots, n)$$

can be arranged in columns to form a matrix solution $M(x)$ and we obtain e^{xA} by calculating $M(x)M(0)^{-1}$.

The solution with initial value v_k is $e^{xA}v_k$ and occupies the kth column of $M(x)$. This seems to require prior knowledge of e^{xA}. However, the vectors v_k can be chosen so that the solutions have an easily computable form.

This is where knowledge of the eigenvalues enters the picture. Let the *distinct* eigenvalues be $\lambda_1, \dots, \lambda_p$. We work over the field $\mathbb{C}$, thus admitting complex vectors in $\mathbb{C}^n$. This may be necessary, since, though A is real, there may be complex eigenvalues.

We assume some facts from linear algebra. For each eigenvalue λ_k of A, the sequence of vector subspaces

$$W_k^m = \ker(\lambda_k I_n - A)^m, \quad m = 0, 1, 2, \ldots,$$

is increasing (ordered by set inclusion), but they cannot grow in dimension indefinitely; hence there exists a lowest positive integer s_k, such that $W_k^{s_k} = W_k^{s_k+1}$. It is known from linear algebra that $\mathbb{C}^n$ is the direct sum

$$\mathbb{C}^n = W_1^{s_1} \oplus \cdots \oplus W_p^{s_p} \tag{5.3}$$

of the spaces $W_k^{s_k}$, $(k = 1, \ldots, p)$. Therefore we can build a basis for $\mathbb{C}^n$ by finding a basis for each of the spaces $W_k^{s_k}$, and uniting them into one set of vectors. This is the basis we shall use to construct the matrix solution $M(x)$. It is useful to recall from linear algebra that the dimension of $W_k^{s_k}$ equals the algebraic multiplicity of the eigenvalue λ_k, that is, its multiplicity as a root of the characteristic equation.

A solution with initial value in $W_k^{s_k}$ has a simple form because the exponential series can be made to terminate. For suppose that $v \in W_k^{s_k}$. Then $(A - \lambda_k I_n)^j v = 0$ for all $j \geq s_k$, so that we have the following finite expression, where the first equality is left for the reader to verify:

$$e^{xA} v = e^{\lambda_k x} e^{x(A - \lambda_k I_n)} v = e^{\lambda_k x} \sum_{j=1}^{s_k - 1} \frac{1}{j!} x^j (A - \lambda_k I_n)^j v.$$

Example Calculate e^{xA} for the matrix:

$$A = \begin{bmatrix} 3 & -1 & 1 \\ 2 & 0 & 1 \\ 1 & -1 & 2 \end{bmatrix}$$

We leave most of the actual calculations for the reader to fill in. Firstly, the characteristic polynomial is

$$\det(\lambda I - A) = \lambda^3 - 5\lambda^2 + 8\lambda - 4 = (\lambda - 1)(\lambda - 2)^2.$$

Hence the distinct eigenvalues are 1 and 2, the latter having algebraic multiplicity 2. We enumerate the eigenvalues: $\lambda_1 = 1$, $\lambda_2 = 2$. We next compute the eigenvectors. There is one independent eigenvector belonging to λ_1, for example

$$v_1 = \begin{bmatrix} 0 \\ 1 \\ 1 \end{bmatrix}$$

which spans W_1^1 (and $s_1 = 1$). There is only one independent eigenvector (although the multiplicity is 2) belonging to λ_2, for example

$$v_2 = \begin{bmatrix} 1 \\ 1 \\ 0 \end{bmatrix}$$

Since the multiplicity is 2 we have $s_2 = 2$ and we seek a vector v_3 that satisfies

$$(A - \lambda_2 I)v_3 = v_2.$$

Here we can use

$$v_3 = \begin{bmatrix} 0 \\ 0 \\ 1 \end{bmatrix}$$

The vectors v_2 and v_3 form a basis for W_2^2, whilst v_1, v_2 and v_3 provide a suitable basis for $\mathbb{R}^3$. The solutions beginning at v_1, v_2 and v_3 are, respectively:

$$e^{xA}v_1 = e^x v_1, \quad e^{xA}v_2 = e^{2x}v_2,$$
$$e^{xA}v_3 = e^{2x}\big(I + x(A - 2I)\big)v_3 = e^{2x}(v_3 + xv_2).$$

Therefore

$$M(x) = \begin{bmatrix} 0 & e^{2x} & xe^{2x} \\ e^x & e^{2x} & xe^{2x} \\ e^x & 0 & e^{2x} \end{bmatrix}$$

Finally

$$e^{xA} = M(x)M(0)^{-1} = \begin{bmatrix} (1+x)e^{2x} & -xe^{2x} & xe^{2x} \\ -e^x + (1+x)e^{2x} & e^x - xe^{2x} & xe^{2x} \\ -e^x + e^{2x} & e^x - e^{2x} & e^{2x} \end{bmatrix}$$

The decomposition (5.3) can be used to prove the exponential decay as $x \to \infty$ of solutions of $y' = Ay$, given that all eigenvalues have negative real parts. The result, in particular, asserts the *stability* of the zero solution. The subject of stability will be taken up in Sect. 6.3 and this result will play a key role. In the proof we briefly need a *matrix norm*. This is a norm on the vector space of matrices, such that $|A^j x| \leq \|A\|^j \, |x|$ for all vectors x, matrices A and positive powers j.

Proposition 5.3 *Suppose that all eigenvalues of the matrix A have negative real parts. Then there exist $C > 0$ and $m > 0$, such that*

$$|e^{xA}\eta| \leq Ce^{-mx}|\eta|, \quad (x > 0)$$

for all $\eta \in \mathbb{R}^n$.

Proof We prove the inequality for all $\eta \in \mathbb{C}^n$. We use the Hermitian norm, assigning the vector $\eta = (z_1, \ldots, z_n) \in \mathbb{C}^n$ the norm

$$|\eta| = (|z_1^2 + \cdots + |z_n|^2)^{1/2}.$$

Recall the decomposition, given by the direct sum (5.3):

$$\eta = v_1 + \cdots + v_p$$

where $v_k \in W_k^{s_k}$. There exists a constant $L > 0$, independent of η and k, such that $|v_k| \leq L|\eta|$ (this is because the projection mapping η to v_k is continuous). We define $\sigma > 0$, such that $-\sigma = \max_k \operatorname{Re} \lambda_k$, and we define $m > 0$, such that $-\sigma < -m < 0$. We first prove that the required inequality holds, with some constant C_k and the stated m, for $\eta = v \in W_k^{s_k}$.

Let $v \in W_k^{s_k}$. Then

$$e^{xA}v = e^{\lambda_k x}e^{(A-\lambda_k I)x}v = e^{\lambda_k x}\sum_{j=0}^{s_k-1}\frac{1}{j!}(A-\lambda_k I)^j x^j v.$$

Hence

$$|e^{xA}v| \leq e^{(\operatorname{Re}\lambda_k)x}\sum_{j=0}^{s_k-1}\frac{1}{j!}\|A-\lambda_k I\|^j |x|^j |v| \leq e^{-\sigma x}P_k(|x|)|v|.$$

where P_k is a polynomial of degree $s_k - 1$, depending only on A and the eigenvalue λ_k. Now there exists $C_k > 0$, such that

$$e^{-\sigma x}P_k(|x|) \leq C_k e^{-mx} \quad (x > 0).$$

Let $\eta \in \mathbb{C}^n$ and let $\eta = \sum_{k=1}^{p} v_k$ with $v_k \in W_k^{s_k}$. Then

$$|e^{xA}\eta| \leq \sum_{k=1}^{p}|e^{xA}v_k| \leq \sum_{k=1}^{p}C_k e^{-mx}|v_k| \leq Ce^{-mx}|\eta|$$

for $x > 0$, where $C = L\max_k C_k$. □

5.2.2 *Cayley-Hamilton*

The characteristic polynomial of the $n \times n$-matrix A is the nth degree polynomial $P(X) = \det(XI_n - A)$. Its leading term is X^n. The roots of $P(X)$ are the eigenvalues of A and the multiplicity of an eigenvalue λ is the multiplicity of λ as a root of $P(X) = 0$; that is, it is the highest power of $X - \lambda$ that divides $P(X)$.

In general if $F(X)$ is the polynomial

$$F(X) = a_m X^m + a_{m-1} X^{m-1} + \cdots + a_1 X + a_0$$

then for a given $n \times n$ matrix A we define

$$F(A) = a_m A^m + a_{m-1} A^{m-1} + \cdots + a_1 A + a_0 I_n.$$

The correspondence $F(X) \mapsto F(A)$ is a homomorphism from the ring of polynomials in the indeterminate X to the ring of $n \times n$ matrices.

The Cayley-Hamilton theorem (proved in courses on algebra) says that if $P(X)$ is the characteristic polynomial of A then $P(A) = 0$. As a result, every power A^k with $k \geq n$ can be expressed as a polynomial in A of degree less than or equal to $n - 1$. Thus

$$e^{xA} = \sum_{k=0}^{\infty} \frac{1}{k!} x^k A^k = \sum_{k=0}^{n-1} Q_k(x) A^k$$

for certain functions $Q_k(x)$, $(k = 0, 1, \ldots, n - 1)$, which clearly depend only on the characteristic polynomial of A, and therefore only on the eigenvalues and their multiplicities.

Calculating the matrix e^{xA} therefore reduces to calculating the functions $Q_k(x)$, and it should be possible to carry this out using only the eigenvalues and their multiplicities. A change of basis for the space of polynomials of degree less than n greatly simplifies this problem. Let the eigenvalues be arranged as a sequence $(\lambda_1, \lambda_2, \ldots, \lambda_n)$ where an eigenvalue with multiplicity r occurs r times in a row. A new basis for the space of polynomials of degree less than n, replacing the powers $1, X, \ldots, X^{n-1}$, is supplied by the sequence $M_k(X)$, $(k = 0, \ldots, n - 1)$, where

$$M_0(X) = 1, \quad M_k(X) = M_{k-1}(X)(X - \lambda_k), \quad (k = 1, \ldots, n - 1)$$

Taking the next step in the sequence we obtain $M_n(X)$, but this is simply the characteristic polynomial of A. By Cayley-Hamilton, $M_n(A) = 0$.

Now we can express e^{xA} in the form

$$e^{xA} = \sum_{k=0}^{n-1} p_k(x) M_k(A) \tag{5.4}$$

for certain functions $p_k(x)$. It turns out that these can be found by a simple algorithm.

Proposition 5.4 (Putzer's Algorithm) *Suppose that the functions $p_k(x)$ satisfy the initial value problems*

$$p_0'(x) - \lambda_1 p_0(x) = 0, \quad p_0(0) = 1$$

$$p_k'(x) - \lambda_{k+1} p_k(x) = p_{k-1}(x), \quad p_k(0) = 0, \quad k = 1, \dots, n-1.$$

Then

$$e^{xA} = \sum_{k=0}^{n-1} p_k(x) M_k(A).$$

Proof Let $T(x) = \sum_{k=0}^{n-1} p_k(x) M_k(A)$, where the functions $p_k(x)$ satisfy the initial value problems as shown. We have (note the telescoping series and the concealed use of Cayley-Hamilton at the final step):

$$\begin{aligned}
T'(x) &= \lambda_1 p_0(x) + \sum_{k=1}^{n-1} \Big(\lambda_{k+1} p_k(x) + p_{k-1}(x)\Big) M_k(A) \\
&= \lambda_1 p_0(x) + \sum_{k=1}^{n-1} \lambda_{k+1} p_k(x) M_k(A) + \sum_{k=1}^{n-1} p_{k-1}(x) M_{k-1}(A)(A - \lambda_k) \\
&= \lambda_1 p_0(x) + \sum_{k=1}^{n-1} \Big(\lambda_{k+1} p_k(x) M_k(A) - \lambda_k p_{k-1}(x) M_{k-1}(A)\Big) + A \sum_{k=1}^{n-1} p_{k-1}(x) M_{k-1}(A) \\
&= \lambda_1 p_0(x) + \lambda_n p_{n-1}(x) M_{n-1}(A) - \lambda_1 p_0(x) + A \sum_{k=1}^{n-1} p_{k-1}(x) M_{k-1}(A) = AT(x).
\end{aligned}$$

Because the initial conditions satisfied by the functions $p_k(x)$ guarantee that $T(0) = M_0(A) = I_n$, we conclude that $T(x) = e^{xA}$. □

It is clear that all the functions $p_k(x)$ are sums of exponential polynomials, with exponents drawn from the set of eigenvalues. By the first equation, $p_0(x) = e^{\lambda_1 x}$. After that it is a question of solving a number of initial value problems of the form

$$y' - \lambda y = x^m e^{\mu x}, \quad y(0) = 0.$$

These can all be handled by the methods of Chap. 1 Sect. 1.4.

Exercise Show that if $\lambda \neq \mu$ the solution to this problem is

$$y(x) = -\frac{m!\, e^{\mu x}}{(\lambda - \mu)^{m+1}} \left(\sum_{k=0}^{m} \frac{(\lambda - \mu)^k x^k}{k!} - e^{(\lambda - \mu)x} \right)$$

whereas if $\lambda = \mu$ it is

$$y(x) = \frac{x^{m+1} e^{\lambda x}}{m+1}.$$

Example Calculate e^{xA} for the matrix (the same as in the previous section):

$$A = \begin{bmatrix} 3 & -1 & 1 \\ 2 & 0 & 1 \\ 1 & -1 & 2 \end{bmatrix}$$

The characteristic polynomial is $(\lambda - 1)(\lambda - 2)^2$. We enumerate the eigenvalues *with repetition*:

$$\lambda_1 = 1, \quad \lambda_2 = 2, \quad \lambda_3 = 2$$

Putzer's algorithm (the reader should solve the corresponding differential equations) now gives

$$p_0(x) = e^x, \quad p_1(x) = e^{2x} - e^x, \quad p_2(x) = (x-1)e^{2x} + e^x$$

and

$$e^{xA} = e^x I + (e^{2x} - e^x)(A - I) + \big((x-1)e^{2x} + e^x\big)(A - I)(A - 2I),$$

a formula that only depends on the sequence of eigenvalues and their multiplicities. On multiplying out the matrices we obtain the same result as before, as the reader may check.

5.2.3 *Interpolation Polynomials*

In the previous section we derived a polynomial

$$R_x(X) := \sum_{k=0}^{n-1} p_k(x) M_k(X)$$

in the indeterminate X, of degree less than or equal to $n - 1$, that satisfied $e^{xA} = R_x(A)$, the coefficients being obtained by solving the differential equations prescribed in Putzer's algorithm. There are other, purely algebraic ways, to obtain a polynomial that yields e^{xA} when A replaces X, and they may prove a better option than Putzer's algorithm.

We are going to work with two pictures of the eigenvalues of A as a sequence. Firstly, as in the previous section, we exhibit them as a sequence of n terms, $(\lambda_1, \ldots, \lambda_n)$, in which each eigenvalue is repeated according to its multiplicity. Secondly we work with the sequence of *distinct* eigenvalues $(\mu_1, \ldots, \mu_p)$, and assign to the eigenvalue μ_k its multiplicity r_k.

For the methods we shall describe we will need to find a polynomial $S_x(X)$ of degree $n-1$ in the indeterminate X, depending on a parameter x, such that, at each eigenvalue μ_k (using the picture without repetitions where μ_k is assigned the multiplicity r_k), the polynomial $S_x(X)$ agrees, as a function of X, with e^{xX} to order r_k. The meaning attached to this is as follows: for each k in the range $1 \leq k \leq p$, the difference $S_x(X) - e^{xX}$ is 0 at $X = \mu_k$, as are all its derivatives with respect to X up to that of order $r_k - 1$, or, in symbols:

$$\frac{d^j S_x}{dX^j}(\mu_k) = x^j e^{x\mu_k}, \quad j = 0, \ldots, r_k - 1, \quad k = 1, \ldots, p.$$

If $S_x(X)$ has the property described in the last paragraph, then one can show by complex analysis, that $e^{xA} = S_x(A)$. A proof of this is outlined below. The point of this is, that the polynomial $S_x(X)$, which is rather obviously unique, can be constructed to agree in this way with e^{xX}, using a purely algebraic algorithm, thus obviating the need to solve differential equations, such as appear in Putzer's algorithm.

Exercise Prove that the polynomial $S_x(X)$ having the stated properties is unique.

In view of the foregoing description of $S_x(X)$, from now on we take the indeterminate X to be a complex variable z.

5.2.4 Newton's Divided Differences

The calculation of e^{xA} can be summarised from the previous section as follows:

Compute the polynomial $S_x(z)$ of degree $n-1$ in z, that, for each k, agrees with e^{xz} at $z = \mu_k$ to order r_k. Then $e^{xA} = S_x(A)$.

A polynomial $S(z)$ (or $S_x(z)$ where a parameter is involved, as for e^{xz}), that agrees with a given analytic function $F(z)$ to order r_k at each point μ_k, $(k = 1, \ldots, p)$ can be found by solving a system of n linear equations to determine its coefficients. But the usefulness of the method lies in the fact that there are also "pencil and paper" algorithms for computing $S(z)$ that bypass the solving of linear equations. These are described in treatises of numerical analysis under the heading of interpolation polynomials.

The most familiar cases are the extreme ones: the case when there is only one point μ, which therefore has multiplicity n; and the case when there are n points, so that all the multiplicities are 1. The first case produces the Taylor polynomial of

$F(z)$ with centre μ and degree $n-1$; the second can be solved by the Lagrange interpolation formula, or else by Newton's method of divided differences computed from the points μ_k. In both cases (using Newton for the second) we can write the solution in the form

$$S(z) = \sum_{k=0}^{n-1} q_k M_k(z)$$

using the polynomial basis $M_k(z)$, $(k = 0, 1, \ldots, n-1)$ introduced in the previous section. We now give a fuller description of these two cases, followed by the general case:

Case A. The matrix A has only one eigenvalue μ, which therefore has multiplicity n. We have $M_k(z) = (z-\mu)^k$ and

$$S(z) = \sum_{k=0}^{n-1} \frac{F^{(k)}(\mu)}{k!}(z-\mu)^k.$$

Hence, applying this to $F(z) = e^{xz}$, we have

$$e^{xA} = e^{x\mu} \sum_{k=0}^{n-1} \frac{x^k}{k!}(A - \mu I_n)^k.$$

Case B. The matrix A has n distinct eigenvalues $\mu_1, \ldots, \mu_n$, which therefore all have multiplicity 1. In this case there is no difference between the sequences $(\mu_1, \ldots, \mu_n)$ and $(\lambda_1, \ldots, \lambda_n)$ and for reasons which will become clear we prefer the latter notation. Then

$$M_0(z) = 1, \quad M_k(z) = (z-\lambda_1)\ldots(z-\lambda_k), \quad (k = 1, \ldots, n-1)$$

and

$$S(z) = \sum_{k=0}^{n-1} F[\lambda_1, \ldots, \lambda_{k+1}] M_k(z).$$

The coefficients here are the *Newton divided differences* for the data points. They are constructed according to the following scheme, for which fuller explanations can be sought in treatises on polynomial approximation:

$$
\begin{array}{llllll}
\lambda_1 & F[\lambda_1] & & & & \\
 & & F[\lambda_1, \lambda_2] & & & \\
\lambda_2 & F[\lambda_2] & & F[\lambda_1, \lambda_2, \lambda_3] & & \\
 & & F[\lambda_2, \lambda_3] & & F[\lambda_1, \lambda_2, \lambda_3, \lambda_4] & \\
\lambda_3 & F[\lambda_3] & & F[\lambda_2, \lambda_3, \lambda_4] & & F[\lambda_1, \lambda_2, \lambda_3, \lambda_4, \lambda_5] \\
 & & F[\lambda_3, \lambda_4] & & F[\lambda_2, \lambda_3, \lambda_4, \lambda_5] & \\
\lambda_4 & F[\lambda_4] & & F[\lambda_3, \lambda_4, \lambda_5] & & F[\lambda_2, \lambda_3, \lambda_4, \lambda_5, \lambda_6] \\
\vdots & \vdots & \vdots & \vdots & \vdots & \vdots \\
\vdots & \vdots & \vdots & \vdots & \vdots & \vdots
\end{array}
$$

The entries can be defined inductively by the divided difference rule:

$$F[\lambda_\ell] = F(\lambda_\ell), \quad F[\lambda_\ell, \ldots, \lambda_{\ell+k}] = \frac{F[\lambda_{\ell+1}, \ldots, \lambda_{\ell+k}] - F[\lambda_\ell, \ldots, \lambda_{\ell+k-1}]}{\lambda_{\ell+k} - \lambda_\ell}$$

A completely different way to construct this polynomial in case B, is to use Lagrange interpolation:

$$\sum_{k=1}^{n} F(\lambda_k) L_k(z), \quad L_k(z) = \prod_{\substack{j=1 \\ j \neq k}}^{n} (z - \lambda_j) \Big/ \prod_{\substack{j=1 \\ j \neq k}}^{n} (\lambda_k - \lambda_j)$$

Which method to use is a question of personal preference. Lagrange interpolation treats the eigenvalues in an egalitarian fashion. The main advantage of the Newton scheme is that one can add further points without having to recalculate the whole table.

Case C: *The General Case.* Newton's divided differences can be generalised to the case that some eigenvalues have multiplicities greater than 1. The method is sometimes known as Hermite interpolation. We suppose that the eigenvalues are λ_1, $\ldots$, λ_n, arranged *in increasing order*, and each eigenvalue is repeated as many times as its multiplicity. The formula for $S(z)$ is the same as in case B, but the definition of $F[\lambda_\ell, \ldots, \lambda_{\ell+k}]$ needs to be amended if $\lambda_\ell = \cdots = \lambda_{\ell+k}$ by defining

$$F[\lambda_\ell, \ldots, \lambda_{\ell+k}] = \frac{1}{k!} F^{(k)}(\lambda_\ell).$$

We give three illustrations, each with three points. Firstly, with distinct points:

$$
\begin{array}{llll}
\lambda_1 & F(\lambda_1) & & \\
 & & \frac{F(\lambda_2)-F(\lambda_1)}{\lambda_2-\lambda_1} & \\
\lambda_2 & F(\lambda_2) & & \frac{\frac{F(\lambda_3)-F(\lambda_2)}{\lambda_3-\lambda_2}-\frac{F(\lambda_2)-F(\lambda_1)}{\lambda_2-\lambda_1}}{\lambda_3-\lambda_1} \\
 & & \frac{F(\lambda_3)-F(\lambda_2)}{\lambda_3-\lambda_2} & \\
\lambda_3 & F(\lambda_3) & &
\end{array}
$$

Secondly, with $\mu = \lambda_1 = \lambda_2 < \lambda_3$:

$$
\begin{array}{llll}
\lambda_1 & F(\mu) & & \\
 & & F'(\mu) & \\
\lambda_2 & F(\mu) & & \frac{\frac{F(\lambda_3)-F(\mu)}{\lambda_3-\mu}-F'(\mu)}{\lambda_3-\mu} \\
 & & \frac{F(\lambda_3)-F(\mu)}{\lambda_3-\mu} & \\
\lambda_3 & F(\lambda_3) & &
\end{array}
$$

Finally, with all three points equal to μ:

$$
\begin{array}{llll}
\lambda_1 & F(\mu) & & \\
 & & F'(\mu) & \\
\lambda_2 & F(\mu) & & \frac{1}{2}F''(\mu) \\
 & & F'(\mu) & \\
\lambda_3 & F(\mu) & &
\end{array}
$$

Example Calculate e^{xA} for the matrix

$$
A = \begin{bmatrix} 3 & -1 & 1 \\ 2 & 0 & 1 \\ 1 & -1 & 2 \end{bmatrix}
$$

The characteristic polynomial is $(\lambda-1)(\lambda-2)^2$. We enumerate the eigenvalues *with repetition*:

$$
\lambda_1 = 1, \quad \lambda_2 = 2, \quad \lambda_3 = 2
$$

The table of divided differences for e^{xz} (as a function of z) is

$$
\begin{array}{llll}
1 & e^x & & \\
 & & e^{2x} - e^x & \\
2 & e^{2x} & & xe^{2x} - e^{2x} + e^x \\
 & & xe^{2x} & \\
2 & e^{2x} & &
\end{array}
$$

The first column shows the eigenvalues. The second entry in the third column is the derivative of e^{xz} with respect to z at $z = 2$; other entries are divided differences. The result, read from the entries along the top diagonal, is

$$e^{xA} = e^x I + (e^{2x} - e^x)(A - I) + \big((x-1)e^{2x} + e^x\big)(A - I)(A - 2I).$$

This is the same formula as we obtained from Putzer's algorithm, but this time the computational effort is arguably less.

5.2.5 *Analytic Functions of a Matrix*

It is useful to put the preceding considerations into the wider context of *analytic functions of a matrix*. This section provides proof of a crucial claim in the previous section; however, the reader lacking background in complex analysis might omit it.

We saw in Sect. 5.2.2 how to define a polynomial function of an $n \times n$-matrix A; essentially, when $F(X)$ is a polynomial we define $F(A)$ by replacing X by A and 1 by I_n. We can extend this to complex analytic functions, but the substitution of X by A no longer has an obvious meaning. All the same, for a class of complex analytic functions $F(z)$, including all entire functions, it is possible to define $F(A)$. This includes the case $F(z) = e^z$ and the resulting $F(A)$ is the same exponential e^A that we have previously encountered.

In what follows we shall assume some knowledge of complex analysis, in particular meromorphic functions and Cauchy's theorem. We shall apply these to *matrix valued functions of the complex variable* z. Integrals of such functions are defined in the obvious way, by integrating each entry of the matrix.

Let A be a matrix and let $F(z)$ be a function of a complex variable, analytic in an open set including the spectrum (that is, the set of all eigenvalues) of A. Then the matrix $F(A)$ is defined by the contour integral

$$F(A) = \frac{1}{2\pi i} \int_C F(z)(zI_n - A)^{-1}\, dz,$$

where the contour C is chosen so that F is analytic in a neighbourhood of C and in the interior domain of C, and so that the winding number of C with respect to each point in the spectrum of A is $+1$. Intuitively, C is supposed to turn once around each eigenvalue in the positive direction. Note that $F(A)$ will in general be a complex matrix, even though A is real.

The correspondence $F \mapsto F(A)$ is a *ring homomorphism*, that maps the function 1 to the unit matrix I_n, and maps z to A. It is often called the functional calculus for the operator A. We will accept these facts, available in linear algebra texts, without proof. Thus if $F(z)$ and $G(z)$ are functions each analytic in some open set containing the eigenvalues of A, then

$$(FG)(A) = F(A)G(A), \quad (F+G)(A) = F(A) + G(A).$$

It follows that the definition of $F(A)$ extends the notion of a polynomial of the matrix A, as defined earlier in Sect. 5.2.2, to the notion of analytic function $F(A)$ of a matrix A, provided $F(z)$ is analytic in an open set including all eigenvalues of A. It also follows that if $F(z)$ and $G(z)$ are two complex analytic functions having the properties just stated, then $F(A)$ and $G(A)$ commute.

According to the previous paragraph the representation of e^{xA} should be the contour integral

$$e^{xA} = \frac{1}{2\pi i}\int_C e^{xz}(zI_n - A)^{-1}\,dz$$

and we can take C to be a positively oriented circle enclosing all eigenvalues. It is not hard to show that this is consistent with the exponential function that we have encountered previously.

Exercise Basic to studies of the *resolvent function* $(zI_n - A)^{-1}$ is the Neumann expansion, actually its Laurent series at $z = \infty$:

$$(zI_n - A)^{-1} = \sum_{k=0}^{\infty} z^{-k-1}A^k,$$

valid for $|z| > \max\{|\lambda| : \lambda \text{ an eigenvalue of } A\}$. Use it to show that:

(a) If $F(z) = z^m$ then $F(A) = A^m$, $(m = 0, 1, 2, \ldots)$. Deduce that if $F(z)$ is a polynomial then the two definitions of $F(A)$, that given in Sect. 5.2.2 and that given by the contour integral, coincide.
(b) If $F_x(z) = e^{xz}$, then $(d/dx)F_x(A) = AF_x(A)$. Deduce that the two definitions given for e^{xA}, that given by the matrix equation $dT/dx = AT$ and that given by the contour integral, coincide.

The functional calculus for the operator A allows an easy proof of the so-called spectral mapping theorem. Our interest here is in the special case of finding the eigenvalues of e^A when the eigenvalues of A are known.

Proposition 5.5 *The eigenvalues of e^A consist of all e^λ where λ ranges over the eigenvalues of A.*

Proof Firstly

$$e^z - e^\lambda = G(z)(z - \lambda)$$

where $G(z)$ is an entire analytic function. Hence

$$e^A - e^\lambda I_n = G(A)(A - \lambda I_n)$$

If λ is an eigenvalue of A we let η be an eigenvector. Then

$$(e^A - e^\lambda I_n)\eta = G(A)(A - \lambda I_n)\eta = 0$$

so that η is also an eigenvector for e^A with eigenvalue e^λ.

Conversely, suppose that $\mu \neq e^\lambda$, for any eigenvalue λ of A. Then the function $H(z) := (e^z - \mu)^{-1}$ is well-defined and analytic in some open set containing the eigenvalues of A. By the operational calculus we have

$$H(A)(e^A - \mu I_n) = (e^A - \mu I_n)H(A) = I_n$$

so that $e^A - \mu I_n$ is invertible; in other words μ is not an eigenvalue of e^A. □

Exercise Is it true that if λ is not an eigenvalue of A then e^λ is not an eigenvalue of e^A?

It turns out that the matrix $F(A)$ can be computed without working out any integral. Quite generally, given an analytic function $F(z)$ and a sequence $(\mu_1, \ldots, \mu_p)$ of distinct complex numbers, such that μ_k is assigned a multiplicity $r_k \geq 1$, we can construct the polynomial $S(z)$ of degree less than n, where $n = \sum_{k=1}^p r_k$, that agrees with $F(z)$ at each μ_k to order r_k. The polynomial $S(z)$ can be constructed by the method of divided differences, as explained in the previous section. If we take the sequence $(\mu_1, \ldots, \mu_p)$ to be the eigenvalues of A, and the numbers r_k their multiplicities, it turns out that the polynomial $S(z)$ has the property that $F(A) = S(A)$. This is how we computed e^{xA} in the previous section.

We shall outline the proof that $F(A) = S(A)$. The key is to note that the matrix function $(zI_n - A)^{-1}$ is an analytic function of z, except for poles at the eigenvalues, and the order of the pole at the eigenvalue μ_k is at most r_k (although it can be lower). This can be seen from the expression

$$(zI_n - A)^{-1} = \frac{1}{\det(zI_n - A)} \operatorname{adj}(zI_n - A)$$

where adj (B) denotes the adjugate matrix of B, that is, the transpose of the matrix of cofactors of B. This formula (in essence Cramer's rule for inverting $zI_n - A$) is the basis of one of the proofs of the Cayley-Hamilton theorem.

We therefore have

$$\begin{aligned}F(A) &= \frac{1}{2\pi i}\int_C F(z)(zI_n - A)^{-1}\,dz \\ &= \frac{1}{2\pi i}\int_C \big(F(z) - S(z)\big)(zI_n - A)^{-1}\,dz + \frac{1}{2\pi i}\int_C S(z)(zI_n - A)^{-1}\,dz \\ &= S(A)\end{aligned}$$

since, in the first integral in the second line the poles of $(zI_n - A)^{-1}$ are cancelled out by the zeros of $F(z) - S(z)$, so that the integral is 0 by Cauchy's theorem.

5.2.6 Exercises

1. Calculate e^{xA} in the following cases, using any convenient method. You can even try more than one method and see which you prefer.

   (a) $A = \begin{bmatrix} 5 & 1 \\ -4 & 0 \end{bmatrix}$ (Eigenvalues $1, 4$)

   (b) $A = \begin{bmatrix} 2 & -1 \\ 1 & 2 \end{bmatrix}$ (Eigenvalues $2 + i, 2 - i$)

   (c) $A = \begin{bmatrix} 0 & 1 \\ 1 & -1 \end{bmatrix}$ (Eigenvalues $-\frac{1}{2} + \frac{1}{2}\sqrt{5}, -\frac{1}{2} - \frac{1}{2}\sqrt{5}$)

   (d) $A = \begin{bmatrix} 5 & -3 \\ 3 & -1 \end{bmatrix}$ (Eigenvalues $2, 2$)

2. Calculate e^{xA} in the following cases. Again you can compare different methods.

   (a) $A = \begin{bmatrix} -5 & 4 & 3 \\ -3 & 2 & 3 \\ -4 & 4 & 2 \end{bmatrix}$ (Eigenvalues $-1, 2, -2$)

   (b) $A = \begin{bmatrix} -1 & 1 & 1 \\ 2 & 1 & 1 \\ 0 & 2 & 1 \end{bmatrix}$ (Eigenvalues $-1, -1, 3$)

   (c) $A = \begin{bmatrix} 0 & 0 & 1 \\ 1 & 0 & 0 \\ -11 & -6 & -6 \end{bmatrix}$ (Eigenvalues $-1, -2, -3$)

   (d) $A = \begin{bmatrix} 1 & -1 & 0 \\ 1 & 3 & 0 \\ 1 & 1 & 2 \end{bmatrix}$ (Eigenvalues $2, 2, 2$)

3. If the 2×2 matrix A has distinct eigenvalues λ_1 and λ_2, show that

$$e^{xA} = e^{\lambda_1 x}\left(\frac{A-\lambda_2}{\lambda_1-\lambda_2}\right) + e^{\lambda_2 x}\left(\frac{A-\lambda_1}{\lambda_2-\lambda_1}\right)$$

Note The formula results from using Lagrange interpolation. It has a simple geometric interpretation: the operators in parentheses are the projections to the eigenspaces. This interpretation holds more generally for the formula, obtained by Lagrange interpolation, in the case of n distinct, simple eigenvalues. This produces the formula

$$e^{xA} = \sum_{k=1}^{n} e^{\lambda_k x} L_k(A), \quad L_k(z) = \prod_{\substack{j=1\\ j\neq k}}^{n}(z-\lambda_j) \Big/ \prod_{\substack{j=1\\ j\neq k}}^{n}(\lambda_k-\lambda_j)$$

You can use Cayley-Hamilton to show that $L_k(A)$ is the projection to the null-space of $A - \lambda_k I$.

4. Let A be a real 2×2 matrix and suppose that the eigenvalues of A are $\alpha + i\omega$ and $\alpha - i\omega$ where α and ω are real and $\omega \neq 0$. Show that

$$e^{xA} = e^{\alpha x}\left(\frac{\sin \omega x}{\omega}\right)A + e^{\alpha x}\left(\cos \omega x - \frac{\alpha \sin \omega x}{\omega}\right) I$$

5. Consider the system of equations

$$\frac{dy_1}{dt} = ay_1 + by_2, \quad \frac{dy_2}{dt} = cy_1 + dy_2$$

and suppose that the eigenvalues of the matrix

$$A = \begin{bmatrix} a & b \\ c & d \end{bmatrix}$$

are $\pm i\omega$ on the imaginary axis with $\omega \neq 0$. Here we use t as the independent variable, intuitively thought of as time. This also frees up the letter "x" for use as a new coordinate. We shall identify the *phase curves*, that is, the curves $t \mapsto (y_1(t), y_2(t))$ in the (y_1, y_2)-plane traced by solutions $(y_1(t), y_2(t))$ of the system. A phase curve can be viewed as a motion of the point (y_1, y_2) in time.

(a) Let $u + iv$ be a complex eigenvector belonging to $-i\omega$, where u and v are real 2-vectors. Use u and v as basis vectors in $\mathbb{R}^2$ and write $y = x_1 u + x_2 v$, where $y = (y_1, y_2)$. Show that in the coordinates x_1, x_2 the problem takes the form

$$\frac{dx_1}{dt} = -\omega x_2, \quad \frac{dx_2}{dt} = \omega x_1$$

and that the phase curves are concentric circles, on which the point (x_1, x_2) moves with constant angular velocity ω around the origin.

(b) Reverting to the coordinates y_1, y_2, and viewing u and v as points in the (y_1, y_2)-plane, show that the phase curve beginning at u is an ellipse that passes through u, v, $-u$, $-v$, in that order, and then again through u, at equal time intervals.

(c) Of all the ellipses that pass through u, v, $-u$, $-v$ show that the phase curve is the one that encloses the least area.

Suppose that the $n \times n$ matrix A has distinct eigenvalues $\mu_1, \ldots, \mu_p$, with multiplicities $r_1, \ldots, r_p$. For a given analytic function $F(z)$, the polynomial $S(z)$, that agrees with $F(z)$ at each μ_k to order r_k, may have degree as high as $n-1$. For example, if $F(z)$ is itself a polynomial of degree $n-1$, then perforce, $S(z) = F(z)$.

Using more knowledge about A we may be able to describe a polynomial $T(z)$, such that $F(A) = T(A)$, together with a cap on the degree of $T(z)$, independent of $F(z)$, that is lower than $n-1$. This may be more convenient for calculating $F(A)$.

The *minimum polynomial* of A is the polynomial $f(X)$ of lowest degree and leading coefficient 1, such that $f(A) = 0$. Its roots are the eigenvalues $\mu_1, \ldots,$ μ_p, but their multiplicities $s_1, \ldots, s_p$ can be strictly lower than the corresponding multiplicities $r_1, \ldots, r_p$ of the eigenvalues as roots of the characteristic polynomial $P(X)$.

6. Suppose that $F(z)$ is analytic in an open set including all eigenvalues of the matrix A.

   (a) Let $T(z)$ be a polynomial that agrees with $F(z)$ at each eigenvalue μ_k of A to order s_k, where s_k is the multiplicity of μ_k as a root of the minimum polynomial of A. Show that $T(A) = F(A)$.
   Hint The defining property of $T(z)$ implies that

$$F(z) - T(z) = f(z)G(z)$$

   where $f(z)$ is the minimum poynomial and $G(z)$ is an entire analytic function. Use the homomorphism property of the correspondence $F \mapsto F(A)$.

   The result of item (a) is useful if A is known to be diagonalisable. It is proved in linear algebra that the matrix A is diagonalisable if and only if the minimum polynomial has only simple roots, that is, if $s_k = 1$ for each k. Important classes of matrices that are always diagonalisable include real symmetric matrices and orthogonal matrices.

   (b) Suppose that A is diagonalisable and that $T(z)$ is a polynomial, such that $T(\mu_k) = F(\mu_k)$ at each eigenvalue μ_k. Show that $F(A) = T(A)$.

(c) Calculate e^{xA} where

$$A = \begin{bmatrix} -1 & \sqrt{2} & -1 \\ \sqrt{2} & 0 & -\sqrt{2} \\ -1 & -\sqrt{2} & -1 \end{bmatrix}$$

Hint The eigenvalues are -2, -2, 2, but, being symmetric, A is diagonalisable.

7. The representation of e^{xA} in the form (5.4) is not unique. For example, if $A = I_3$ then $M_1(A) = M_2(A) = 0$; so $p_1(x)$ and $p_2(x)$ can be arbitrarily assigned.

   (a) Show that the representation (5.4) fails to be unique if and only if the minimum polynomial of A has degree strictly less than n.
   (b) Show that the coefficient functions assigned by Putzer's algorithm and those assigned by the method of divided differences are the same.
   Hint They only depend on the eigenvalues and their multiplicities.

Proposition 5.3 was concerned with the decay of solutions of $y' = Ay$ as $x \to \infty$. We can also ask about the existence of solutions bounded on the whole line $\mathbb{R}$. These would include periodic solutions, but also others, with multiple periods, such as $e^{ix} + e^{\sqrt{2}ix}$, that belong to the class of *almost periodic functions*. It should be obvious to the reader that if A has an eigenvalue on the imaginary axis then there exists a bounded solution, not identically zero; in fact a real one, if A is a real matrix. The converse is extremely plausible, that if a bounded, non-identically zero solution exists, then A has an eigenvalue on the imaginary axis. The reader was asked to consider an example of this in 1.4 Exercise 2 (for a fourth order scalar equation). The general case seems tricky to prove. We resolve it by introducing the *Laplace transform*.

Many problems involving linear equations with constant coefficients can be handled by the Laplace transform. It is an important technique, often used in practical mathematics to solve the Cauchy problem for inhomogeneous linear equations with constant coefficients, and we could have presented it in Chap. 1 for this purpose. It is particularly useful when the inhomogeneous term is discontinuous (in Sect. 5.3 project B we shall consider the theory of such equations, including the possibility of discontinuous coefficient functions). We introduce it here to approach a question which otherwise can appear quite tricky.

Very briefly, given a function $f(x)$ (possibly vector valued), which is integrable on the interval $[0, K]$ for all $K > 0$, the Laplace transform of f is the function

$$\tilde{f}(p) := \int_0^\infty f(x)e^{-px}\,dx$$

defined, in the first place, for all *complex* p for which the integral is convergent, but usually extended analytically to a larger domain. We will not develop any general properties of the Laplace transform here, but simply point to some salient facts, asking the reader to prove them in the next exercise. Following that, the reader can apply it to the question of bounded solutions of $y' = Ay$.

8. Prove the following properties of the Laplace transform:

   (a) If f is bounded in $[0, \infty[$ then the integral defining $\tilde{f}(p)$ is convergent for all p, such that $\operatorname{Re} p > 0$.

   (b) If n is a natural number and $f(x) = x^n e^{\lambda x}$ then

   $$\tilde{f}(p) = \frac{n!}{(p - \lambda)^{n+1}}.$$

   Hint Reduce to the case $n = 0$.
   Note The integral defining the Laplace transform here is convergent for $\operatorname{Re} p > \operatorname{Re} \lambda$, but the transform is analytically extended to a meromorphic function of p with a pole at λ.

   (c) If $f(x)$ is an exponential polynomial

   $$f(x) = \sum_{k=1}^{m} g_k(x) e^{\lambda_k x}$$

   with non-zero polynomials $g_k(x)$ and distinct exponents λ_k, $(k = 1, \dots, m)$, then $\tilde{f}(p)$ is a meromorphic function (when analytically extended) with poles at the points λ_k.

9. (a) Show that the problem

   $$p_n y^{(n)} + p_{n-1} y^{(n-1)} + \cdots + p_1 y' + p_0 y = 0$$

   has a bounded solution $y(x)$, not identically zero, if and only if the indicial equation has a root on the imaginary axis.
   Hint If $y(x)$ is a bounded solution consider the Laplace transforms of $y(x)$ and $y(-x)$.

   (b) Show that the vector equation $y' = Ay$ has a non-zero solution that is bounded on the whole real line if and only if A has an eigenvalue on the imaginary axis.

   (c) Suppose that all eigenvalues of A are on the imaginary axis. Does it follow that all non-zero solutions of $y' = Ay$ are bounded?
   Hint If $\eta \in W_k^{s_k}$, and $\lambda_k = i\beta$, the coordinates of the solution $e^{xA}\eta$ are exponential polynomials with exponent $i\beta$ and degree up to $s_k - 1$. A full account must acknowledge the possible difference between the algebraic and the geometric multiplicities of the eigenvalue $i\beta$.

5.2.7 Projects

The two projects in this section require a good understanding of the derivative of a multivariate function as a linear mapping. Feeling comfortable with matrix algebra is also helpful.

A. *Project on the derivative of the determinant function*

We develop a formula for the derivative of the determinant function, regarded as a function from the vector space $M^{n\times n}$ of real $n \times n$ matrices to the reals. The vector space $M^{n\times n}$ has dimension n^2 and therefore has a unique topology derivable from a norm. There are a number of norms commonly used for this matrix space, but we will not need any of them here.

A1. Let $A(x)$ be a matrix valued function of the real variable x. Show that

$$\frac{d}{dx} \det A(x) = \operatorname{tr}\bigl(A(x)^{-1}A'(x)\bigr)\bigl(\det A(x)\bigr)$$

at all points x for which $\det A(x) \neq 0$. Here $\operatorname{tr} T$ denotes the trace of the matrix T, that is, the sum of its eigenvalues counted with multiplicity.
Hint In the case $A(x) = I + xT$, where T is a constant matrix, $\det A(x)$ is the product

$$\det A(x) = \prod_{k=1}^{n} (1 + \lambda_k x)$$

where the numbers $\lambda_1, \ldots, \lambda_n$ are the eigenvalues of T, including complex eigenvalues, repeated according to their multiplicities. Differentiate this formula at $x = 0$. The result can be interpreted as the directional derivative of the function $A \mapsto \det A$, computed at $A = I$ in the "direction" T. For the general case use the chain rule from multivariate calculus.

A2. Prove that $\det e^{xA} = e^{x \operatorname{tr} A}$.
Hint Use Exercise A1 and the fact that, in general, $\operatorname{tr}(ST) = \operatorname{tr}(TS)$.

A3. We consider the restriction of the determinant function to the general linear group $GL(n)$, consisting of all invertible $n \times n$ matrices. This is, of course, the set of matrices A for which $\det(A) \neq 0$, and it is therefore open in $M^{n\times n}$. Show that

$$D(\det)(A)\,T = \operatorname{tr}(A^{-1}T)(\det A), \quad (A \in GL(n),\ T \in M^{n\times n}).$$

The formula expresses the derivative of the function det at the point A in $GL(n)$ as a linear functional, denoted here by $D(\det)(A)$, acting on the vector space $M^{n\times n}$ of all $n \times n$ matrices, and here taking the argument T.

A4. The expression for the derivative of the determinant function at an invertible matrix argument A given in the previous exercise extends to matrix arguments that are not necessarily invertible. The result is called Jacobi's formula:

$$D(\det)(A)\,T = \operatorname{tr}\big((\operatorname{adj} A)T\big), \quad (A, T \in M^{n\times n})$$

where adj A denotes the adjugate matrix of A. Prove this formula.
Hint Observe that $(\det A)A^{-1} = \operatorname{adj} A$ if A is invertible and approximate A by invertible matrices if it is not.
Note The fact that the set of all invertible matrices is dense in the set of all matrices is an easy consequence of the reduction to Jordan canonical form.

B. *Project on the derivative of the exponential function*
We develop a formula for the derivative of the exponential function, regarded as a mapping $\exp : M^{n\times n} \to M^{n\times n}$.

B1. Let A and B be matrices. Verify that the matrix function $e^{xA}Be^{-xA}$ is the solution of the initial value matrix problem

$$\frac{dT}{dx} = AT - TA, \quad T(0) = B.$$

We can define the linear mapping[1] $\operatorname{ad}_A : M^{n\times n} \to M^{n\times n}$ by

$$(\operatorname{ad}_A)\,T = AT - TA.$$

The linear mapping ad_A has its own exponential e^{ad_A}, not an $n \times n$ matrix but a linear mapping from $M^{n\times n}$ to itself (if desired it could be expressed as a matrix with n^4 entries). Exercise B1 established the important formula

$$e^{xA}Te^{-xA} = e^{x\,\operatorname{ad}_A}(T), \quad (T \in M^{n\times n}).$$

B2. In this exercise we find a formula for the derivative of the mapping $\exp : M^{n\times n} \to M^{n\times n}$. Evaluated at a point A this will be a linear mapping

$$D(\exp)(A) : M^{n\times n} \to M^{n\times n},$$

the value of which at $T \in M^{n\times n}$ is denoted by $(D(\exp)(A))\,(T)$. The definition of $\exp A$, or more precisely the formula $(d/dx)e^{xA} = Ae^{xA}$, only tells us what this is when $T = A$, namely, it is the usual matrix product $\exp(A)\,A$.

[1] The notation ad_A is from Lie algebra theory. It is the adjoint representation of A as linear operator acting in the space $M^{n\times n}$.

(a) Let $t \mapsto A(t)$ be a differentiable curve in the space $M^{n\times n}$ and define

$$K(s,t) = e^{-sA(t)}\frac{\partial}{\partial t}\left(e^{sA(t)}\right).$$

Show that

$$\frac{\partial K}{\partial s} = e^{-s\,\mathrm{ad}_{A(t)}}(A'(t)).$$

(b) Show that

$$K(1,t) = \left(\sum_{k=0}^{\infty}\frac{(-1)^k(\mathrm{ad}_{A(t)})^k}{(k+1)!}\right)(A'(t)).$$

Note The series in parentheses is a linear mapping from $M^{n\times n}$ to itself, here taking $A'(t)$ as argument. We indicate this, as in item (a) and the following ones, by the parentheses around the argument.

(c) Define the function $\Phi(z) = (1-e^{-z})/z$. Note that $\Phi(z)$ is an entire function and so $\Phi(S)$ makes sense for any linear mapping S from a finite-dimensional vector space to itself. In particular, given a matrix A, the linear mapping $\Phi(\mathrm{ad}_A)$ maps $M^{n\times n}$ to itself. Show that

$$K(1,t) = \Phi(\mathrm{ad}_{A(t)})(A'(t)).$$

(d) Deduce that

$$(D(\exp)(A))\,(T) = \exp(A)\big(\Phi(\mathrm{ad}_A)(T)\big).$$

Note On the right of this formula we have the usual product of the matrices $\exp(A)$ and $\Phi(\mathrm{ad}_A)(T)$. In the second factor, $\Phi(\mathrm{ad}_A)$ is a linear mapping from $M^{n\times n}$ to itself, here taking T as its argument.

(e) Deduce that if T and A commute then $(D(\exp)(A))\,(T)$ is the ordinary matrix product $\exp(A)\,T$.

5.3 Linear Systems with Variable Coefficients

Let $A(x)$ be an $n \times n$ matrix function of the real variable x. We assume that the entries are continuous functions of x in an interval $]a,b[$. Given x_0 in the interval $]a,b[$ and a vector η, we know that the initial value problem

$$y' = A(x)y, \quad y(x_0) = \eta$$

has a unique solution in the interval $]a, b[$. However, there is no general method for calculating it in closed form.[2] Just as in the case of constant coefficients, the problem $y' = A(x)y$ with variable coefficients is fully solved once we have found n linearly independent solutions (compare the discussion at the beginning of Sect. 5.2.1). Unlike the case of constant coefficients, there is no general way to find such solutions. However, special cases can be solved in closed form. Most of the proofs in this section are left as exercises for the reader.

Let $\phi_1, \ldots, \phi_n$ be n vector solutions of the problem $y' = A(x)y$ and let $a < x_0 < b$. Then the n solutions are linearly independent functions on the interval $]a, b[$ if and only if the vectors $\phi_1(x_0), \ldots, \phi_n(x_0)$ are linearly independent in $\mathbb{R}^n$.

Exercise Prove the claim in the last sentence.

Let $\phi_1, \ldots, \phi_n$ be linearly independent solutions of $y' = A(x)y$ on the interval $]a, b[$. Form the matrix function $M(x)$ by arranging the vectors $\phi_1(x), \ldots, \phi_n(x)$ in its columns, thus:

$$M(x) = \begin{bmatrix} | & \cdots & | \\ \phi_1(x) & \cdots & \phi_n(x) \\ | & \cdots & | \end{bmatrix}$$

Definition A matrix function $M(x)$, whose columns comprise n linearly independent solutions of $y' = A(x)y$, is called a *fundamental matrix*.

Exercise Show that the columns of $M(x)$ are solutions of $y' = A(x)y$ if and only if $M(x)$ satisfies the matrix differential equation $M'(x) = A(x)M(x)$.

Knowing a fundamental matrix we can easily solve the initial value problem. In fact, we used this to calculate e^{xA} by the eigenvector method. Given x_0 in the interval $]a, b[$ and a vector η, the solution of the initial value problem

$$y' = A(x)y, \quad y(x_0) = \eta$$

is

$$y(x) = M(x)M(x_0)^{-1}\eta.$$

Exercise Prove the claim in the last sentence.

This result says that we can solve the initial value problem for arbitrary initial vectors, using only algebraic operations, provided we know n linearly independent solutions of $y' = A(x)y$. We can go further with the same knowledge. We can solve

---

[2] A solution in closed form is generally understood to be one not involving infinite series and built out of elementary transcendental functions. It seems wise to expand the notion to allow recognised special functions (e.g. Bessel functions).

the non-homogeneous equation

$$y' = A(x)y + h(x) \tag{5.5}$$

on the interval $]a, b[$ by quadratures, using the method of variation of parameters. This consists in first seeking a particular solution of (5.5) in the form $y(x) = M(x)v(x)$, where $M(x)$ is a fundamental matrix and $v(x)$ is an unknown vector function.

Exercise Assume that the vector function $h(x)$ is continuous in the interval $]a, b[$. Show that the function $M(x)v(x)$ is a solution of $y' = A(x)y + h(x)$ if and only if $M(x)v'(x) = h(x)$. Hence show that the most general solution of (5.5) is

$$y(x) = M(x)c + M(x)\int M(x)^{-1}h(x)\,dx$$

where c is an arbitrary vector in $\mathbb{R}^n$. The integral is an antiderivative of the vector valued function $M(x)^{-1}h(x)$. The result can be viewed as a generalisation of the solution formula (1.4).

Exercise Show that the solution of the initial value problem

$$y' = A(x)y + h(x), \quad y(x_0) = \eta$$

is given by the formula

$$y(x) = M(x)M(x_0)^{-1}\eta + M(x)\int_{x_0}^{x} M(t)^{-1}h(t)\,dt.$$

In the case that $A(x)$ is a constant matrix A, the result of the exercise reduces to the important and useful formula

$$y(x) = e^{(x-x_0)A}\eta + \int_{x_0}^{x} e^{(x-t)A}h(t)\,dt. \tag{5.6}$$

More generally, for the problem $y' = A(x)y$ we can define the "time advance mapping", a terminology that is sensible if x is time; but we will use it anyway. This is a matrix function of two variables x_0 and x, suggestively denoted by $\Phi^x_{x_0}$, defined so that the solution of the initial value problem

$$y' = A(x)y, \quad y(x_0) = \eta$$

is given by

$$y(x) = \Phi^x_{x_0}\eta.$$

Exercise Show that if $M(x)$ is a fundamental matrix then

$$\Phi_{x_0}^x = M(x)M(x_0)^{-1},$$

and the solution of the non-homogeneous initial value problem is

$$y(x) = \Phi_{x_0}^x \eta + \int_{x_0}^x \Phi_t^x h(t)\, dt.$$

In this form the function Φ_t^x may be viewed as the Green's function for the initial value problem. This rule, which is often referred to as Duhamel's principle, can be extended in scope to solve certain partial differential equations, such as the analogous non-homogeneous problems for the heat equation and wave equation.

Example Find the general solution on the interval $x > 0$ of the problem $y' = A(x)y$, where

$$A(x) = \begin{bmatrix} 1 & x \\ 0 & 1/x \end{bmatrix}$$

and determine the time advance mapping.

Solution In coordinates the problem is

$$y_1' = y_1 + xy_2, \quad y_2' = (1/x)y_2.$$

Solving the second equation for y_2 we obtain $y_2(x) = c_2 x$ where c_2 is an arbitrary constant. Substituting this into the first equation and solving for y_1 we obtain, after a short calculation,

$$y_1(x) = -c_2(x^2 + 2x + 2) + c_1 e^x$$

where c_1 is another arbitrary constant. This yields the general solution and it may be expressed as $y(x) = M(x)c$, where

$$M(x) = \begin{bmatrix} e^x & -x^2 - 2x - 2 \\ 0 & x \end{bmatrix}, \quad c = \begin{bmatrix} c_1 \\ c_2 \end{bmatrix}$$

This indicates that $M(x)$ is a fundamental matrix and a straightforward, though lengthy, matrix calculation leads to

$$\Phi_{x_0}^x = M(x)M(x_0)^{-1} = \begin{bmatrix} e^{x-x_0} & (x_0 + 2 + (2/x_0))e^{x-x_0} - (x + 2 + (2/x))(x/x_0) \\ 0 & x/x_0 \end{bmatrix}$$

where $x > 0$ and $x_0 > 0$.

5.3.1 *Exercises*

1. Obtain a fundamental matrix for the equation $y' = A(x)y$ in the following cases:

$$A(x) = \begin{bmatrix} 1 & x-1 \\ 0 & x \end{bmatrix}, \quad \begin{bmatrix} x & x^2 \\ 0 & x \end{bmatrix}, \quad \begin{bmatrix} 0 & 1 \\ -9/x^2 & -11/x \end{bmatrix} \ (x > 0).$$

   Calculate the time advance mapping $\Phi_{x_0}^x$ for each case.
   Hint For the third case show that the coordinate y_1 satisfies an Euler equation.
2. Let $A(x)$ be a continuous matrix function defined for all real x. Let $\Phi_{x_0}^x$ be the time advance mapping for the equation $y' = A(x)y$. Show that

$$\Phi_a^b = \Phi_c^b \Phi_a^c$$

   for all $a, b, c \in \mathbb{R}$.
3. Find the general solution of the equation $y' = Ay + h(x)$ in the following cases:

   (a) $A = \begin{bmatrix} 5 & 1 \\ -4 & 0 \end{bmatrix}, \qquad h(x) = \begin{bmatrix} 1 \\ 1 \end{bmatrix}$

   (b) $A = \begin{bmatrix} 5 & 1 \\ -4 & 0 \end{bmatrix}, \qquad h(x) = \begin{bmatrix} e^x \\ 0 \end{bmatrix}$

   (c) $A = \begin{bmatrix} 2 & -1 \\ 1 & 2 \end{bmatrix}, \qquad h(x) = \begin{bmatrix} \cos x \\ \sin x \end{bmatrix}$

   Hint See Sect. 5.2 Exercise 1.
4. Find the general solution of the problem $y' = A(x)y + h(x)$ where

$$A(x) = \begin{bmatrix} 0 & 1 \\ -9/x^2 & -11/x \end{bmatrix}, \qquad h(x) = \begin{bmatrix} x \\ -1 \end{bmatrix}$$

   Hint See Exercise 1.
5. Given the matrix function $A(x)$ we can fix a base point x_0 in its interval of definition and form the matrix function

$$M(x) = \exp\left(\int_{x_0}^{x} A(t)\, dt\right).$$

   The analogy with the scalar equation $y' = a(x)y$ suggests that $M(x)$ might be a fundamental matrix for the matrix differential equation $y' = A(x)y$. However, we cannot expect this in general, owing to the fact that matrix multiplication is non-commutative.

   (a) Prove that if $A(x)$ and $\int_{x_0}^{x} A(t)\, dt$ commute for all x, then $M(x)$ is a fundamental matrix for $y' = A(x)y$.
   Hint Consult Sect. 5.2 Exercise A1.

(b) Show that a sufficient condition for $A(x)$ and $\int_{x_0}^{x} A(t)\,dt$ to commute for all x is that

$$A(s)A(t) = A(t)A(s)$$

for all s and t in the interval of definition.

6. Let A be an $(n \times n)$-matrix and let $p(x)$ be continuous in an open interval I. Let $x_0 \in I$. Solve the initial value problem

$$\begin{aligned} y' &= p(x)Ay, \quad (x \in I) \\ y(x_0) &= \eta. \end{aligned}$$

7. Find opportunities to use the results of the previous two exercises in order to obtain a fundamental matrix for $y' = A(x)y$ in the following cases:

$$A(x) = \begin{bmatrix} x & x^2 \\ 0 & x \end{bmatrix}, \quad \begin{bmatrix} x & -x^2 \\ x^2 & x \end{bmatrix}, \quad \begin{bmatrix} 1/x & 1 & 1 \\ 0 & 1/x & 1 \\ 0 & 0 & 1/x \end{bmatrix} \quad (x > 0).$$

8. Let $M(x)$ be a fundamental matrix for the equation $y' = A(x)y$ in the interval I. Let $W(x) = \det M(x)$ and $a(x) = \operatorname{tr} A(x)$. Let $x_0 \in I$. Show that

$$W(x) = W(x_0)e^{\int_{x_0}^{x} a(t)\,dt}, \quad (x \in I).$$

Deduce that $M(x)$ preserves n-dimensional volume in $\mathbb{R}^n$ if and only if $\operatorname{tr} A(x) = 0$ for all $x \in I$.
Hint Consult Sect. 5.2 Exercise A1.
Note The formula for $W(x)$ extends Abel's formula for the Wronskian (see Proposition 1.7). Can you see why?

9. Let A, B and C be $n \times n$-matrices.

(a) Show that the matrix function

$$T(x) = e^{xA}Ce^{xB}$$

satisfies the initial value matrix problem

$$T' = AT + TB, \quad T(0) = C.$$

(b) Suppose that the eigenvalues of A and B all lie in the left half-plane. Show that the matrix

$$X := -\int_0^{\infty} e^{xA}Ce^{xB}\,dx$$

is well-defined, in the sense that the integral is convergent, and that X is the unique matrix that satisfies

$$AX + XB = C.$$

Hint Recall Proposition 5.3.

5.3.2 Projects

A. *Project on periodic differential equations*
We study the linear problem $y' = A(x)y$, where $A(x)$ is a continuous matrix function that is periodic in x with period T. The material covered here is an introduction to what is called *Floquet theory*.

A1. (a) Let $M(x)$ be a fundamental matrix for the problem $y' = A(x)y$. Show that $M(x + T)$ is also a fundamental matrix.
(b) Show that

$$M(x + T) = M(x)M(0)^{-1}M(T)$$

(c) Show that

$$\Phi_0^{x+T} = \Phi_0^x \Phi_0^T$$

for all x, where $\Phi_{x_0}^x$ is the time advance mapping.
(d) Show that for all integers m and n, positive or negative, we have

$$\Phi_0^{mT} \Phi_0^{nT} = \Phi_0^{(m+n)T}.$$

Thus, the matrices Φ_0^{nT}, $(n \in \mathbb{Z})$, form a group isomorphic to $(\mathbb{Z}, +)$.
(e) Deduce that for all integers n we have

$$\Phi_0^{nT+x} = \Phi_0^x (\Phi_0^T)^n.$$

A2. Let $B = \Phi_0^T$. This is called the period advance mapping, or the Poincaré mapping.

(a) Show that the solution of

$$y' = A(x)y, \quad y(0) = \eta$$

is periodic with period T if and only if $B\eta = \eta$.

(b) Let D be a matrix (possibly complex), such that $B = e^{TD}$, and define

$$P(x) = \Phi_0^x e^{-xD}.$$

Show that $P(x)$ is a T-periodic function of x. Note that $P(x)$ could be complex.
Note One way to define D, which is not unique, is by

$$D = \frac{1}{2\pi i T} \int_C (\log z)(zI_n - B)^{-1}\, dz.$$

where a branch of the logarithm is chosen by cutting the complex plane along a ray that avoids all eigenvalues of B, and the contour C lies in the cut plane and encloses all the eigenvalues. See the discussion of analytic functions of a matrix in Sect. 5.2.

(c) Introduce a new dependent variable vector $w \in \mathbb{C}^n$ by setting

$$y = P(x)w.$$

The transformation is T-periodic in x. Show that the problem $y' = A(x)y$ is transformed into the constant coefficient problem

$$w' = Dw.$$

(d) Show that the solution to the initial value problem

$$y' = A(x)y, \quad y(0) = \eta$$

is

$$y(x) = P(x)e^{xD}\eta.$$

The equivalent constant coefficient equation $w' = Dw$ has, in general, complex coefficients. The reason is that the real invertible matrix B does not necessarily have a real logarithm. However, B^2 has a real logarithm; one such is $D + \overline{D}$, as the reader is now asked to prove.

A3. (a) Show that $D\overline{D} = \overline{D}D$.
Hint By the discussion of Sect. 5.2.5 there exists a polynomial $S(z) = \sum_{k=0}^{m} c_k z^k$ with complex coefficients, such that $S(B) = D$.

(b) Deduce that $e^{D+\overline{D}} = B^2$.

It follows that a *real equivalent constant coefficients equation* can be obtained by a coordinate change with period $2T$.

B. *Project on linear equations with discontinuities*
The theory of the linear equation (1.1) developed in Chap. 1, culminating in Proposition 1.10 on the initial value problem, was based on the assumption that the coefficient functions p_j and the inhomogeneous term g were continuous. In applications to technology, initial value problems like (1.11) and boundary value problems like (1.28) occur frequently, in which the function g is discontinuous, and the reader who pursues more applied studies cannot fail to encounter them.

We can consider the equation for damped harmonic oscillations with a forcing term (1.4 Exercise A3). In a practical application a discontinuous forcing term with simple jump discontinuities may be the most natural way to model a situation where, for example, an input signal is switched repeatedly on and off.

Problems in which the coefficient functions have discontinuities also arise naturally. The Schrödinger equation

$$\psi'' + (E - U(x))\psi = 0,$$

considered in 1.4 project B, is often treated with a *square well potential*. For example, we may have $U(x) = -1$ for $-1 < x < 1$, and $U(x) = 0$ for $|x| > 1$.

The correct way to extend the theory of Chap. 1 to such problems, is to slightly weaken the definition of solution, admitting as solutions functions that may, in part, lack derivatives. We first recall from analysis that a function f (which may be vector or matrix valued) defined in a bounded closed interval $[a, b]$ is called *piece-wise continuous* if there is a partition $a = t_0 < t_1 < \cdots < t_m = b$ such that the restriction of f to each open interval $]t_j, t_{j+1}[$ extends to a continuous function in the closed interval $[t_j, t_{j+1}]$. In short, the discontinuities of f are all jumps. A function f defined in an *open interval* I is called piece-wise continuous if it is piece-wise continuous in *every bounded closed interval* $[a, b] \subset I$.

A fairly obvious consequence of the definition is that the discontinuities of a piece-wise continuous function in an open interval I form a discrete set, possibly infinite, with no limit point in I.

We say that a function f is *piece-wise continuously differentiable* if both the following conditions are satisfied:

(i) f is continuous.
(ii) There is a piece-wise continuous function g, such that, at all points x at which g is continuous, f is differentiable and $f'(x) = g(x)$.

We now reconsider the problem (1.1), but allowing the coefficient functions and the inhomogenous term to be piece-wise continuous. In this situation, and for the purposes of this project, we shall say that a function $y(x)$ is a *weak solution* of (1.1) if the following hold:

(i) The function $y(x)$ has continuous derivatives up to order $n - 1$.

(ii) The function $y^{(n-1)}$ is piece-wise continuously differentiable.[3]
(iii) $y(x)$ satisfies the equation for all x at which the coefficient functions and the inhomogeneous term are continuous.

We also reconsider Eq. (5.5), allowing the matrix function $A(x)$ and the inhomogeneous term $h(x)$ to be piece-wise continuous. In this case, and for the purposes of this project, we shall say that a vector function $y(x)$ defined in the interval I is a *weak solution* of $y' = A(x)y + h(x)$ if it satisfies:

(i) $y(x)$ is piece-wise continuously differentiable.
(ii) $y'(x) = A(x)y(x) + h(x)$ at all points where $A(x)$ and $h(x)$ are continuous.

The object of this project is to prove the analogues of Propositions 1.8, 1.9 and 1.10 in the case that the coefficient functions and the inhomogeneous term may be piece-wise continuous. In the following exercises I denotes an open interval.

B1. Let f be a piece-wise continuous function in the interval I. Show that the discontinuities of f can be arranged in a sequence of one of the following types:

(a) An increasing infinite sequence $(t_j)_{j=0}^{\infty}$ without limit point in I.
(b) A decreasing infinite sequence $(t_j)_{j=0}^{\infty}$ without limit point in I.
(c) An increasing bi-infinite sequence $(t_j)_{j=-\infty}^{\infty}$ without limit point in I.
(d) An increasing finite sequence $(t_j)_{j=0}^{m}$.

B2. Let $A(x)$ be a piece-wise continuous $n \times n$ matrix valued function defined in the interval I. Show how to build a continuous matrix valued function $M(x)$, defined in I, such that $M(x)$ is invertible for all x and is a weak solution of the differential equation

$$M'(x) = A(x)M(x).$$

This provides a weak fundamental matrix for $y' = A(x)y$.
Hint Use the previous exercise. Begin by solving the equation in the interval $[t_0, t_1]$ in cases (a), (c) or (d), or in the interval $[t_1, t_0]$ in case (b), using the identity matrix as initial value at the midpoint. Continue to adjacent intervals by matching the endpoint values of $M(x)$.

B3. Let $A(x)$ be a piece-wise continuous matrix function and $h(x)$ a piece-wise continuous vector function, defined in I.

[3] There does not seem to be a suitable name for the class of functions defined here by (i) and (ii). One should avoid calling them "piece-wise C^n", as this would mean a function, possibly discontinuous, that is C^n on each of the intervals of a partition.

(a) Let $M(x)$ be a weak fundamental matrix for $y' = A(x)y$ (see the previous exercise). Show that, for all vectors v, the function

$$y(x) := M(x)v + M(x) \int M(x)^{-1} h(x)\, dx$$

is a weak solution of the differential equation $y' = A(x)y + h(x)$, and, furthermore, every weak solution is of this form with an appropriate value for v.
Note The indefinite integral $\int M(x)^{-1} h(x)\, dx$ denotes a weak solution of the equation $y' = M(x)^{-1} h(x)$. One such is the definite integral

$$\phi(x) := \int_c^x M(t)^{-1} h(t)\, dt$$

where c is a chosen base point in I.
(b) Let $c \in I$ and let $\eta \in \mathbb{R}^n$. Show that the problem

$$y' = A(x)y + h(x), \quad y(c) = \eta$$

has a unique weak solution on the interval I.

B4. Prove the analogues of Propositions 1.8, 1.9 and 1.10 in the case that the coefficient functions and the inhomogeneous term are piece-wise continuous.
Hint Convert to an equivalent first order linear system and use the previous exercise.

B5. Get some practice. Let $p(x) = 1$ for $|x| < \pi$ and $p(x) = 0$ for $|x| > \pi$.

(a) Calculate a solution basis for the equation

$$y'' + p(x)y = 0$$

on the real line $\mathbb{R}$.
(b) Calculate a general solution on $\mathbb{R}$ for the equation

$$y'' + p(x)y = 1.$$

Hint Simplest is to find a particular solution equal to 1 for $|x| < \pi$.

The material of this project is capable of far-reaching generalisation. In particular Caratheodory proved a version of the Picard existence theorem (Proposition 4.1), in which the function $f(x, y)$ is not necessarily continuous in x, but is measurable (a concept the reader may have encountered while studying the Lebesgue integral). The solution obtained is *absolutely continuous* and satisfies the equation *almost everywhere*. The italicised phrases here are also concepts associated with the Lebesgue integral.

The more advanced theory of differential equations is much concerned with weak solutions, but relying on a notion of weak solution, that is more general, more sophisticated and more elegant than that of this project. It is built on a vast expansion of the realm of ordinary functions to include objects, in general not functions, called distributions.

Chapter 6
Continuation of Solutions

> Beauty: it curves, curves are beauty.

We take up again the initial value problem

$$y' = f(x, y), \quad y(x_0) = \eta.$$

We have seen that under certain conditions a unique solution exists on an interval $]x_0 - h, x_0 + h[$ for some h. In this chapter we shall consider whether the solution can be extended, and how far. We also study how the solution depends on the initial conditions, aiming for near optimum regularity results. These have many applications, both theoretical and practical. The chapter ends with an introduction to stability theory.

6.1 The Maximal Solution

We assume that $f = (f_1, \ldots, f_n) : D \to \mathbb{R}^n$, where D is an open, connected subset of $\mathbb{R} \times \mathbb{R}^n$, that f is continuous, and that in addition f satisfies a *local Lipschitz condition* in its second argument, that will enable us to apply Proposition 4.1 locally. Because we will often need this condition we give it a name:

Condition L For every compact $A \subset D$, there exists $K > 0$, such that

$$|f(x, y^{(1)}) - f(x, y^{(2)})| \le K|y^{(1)} - y^{(2)}|$$

for all $(x, y^{(1)})$ and $(x, y^{(2)})$ in A, such that the line segment joining $(x, y^{(1)})$ and $(x, y^{(2)})$ lies in A.

R. Magnus, *Essential Ordinary Differential Equations*, Springer Undergraduate Mathematics Series, https://doi.org/10.1007/978-3-031-11531-8_6

It need not be the case that a constant K exists that works for all of D; that is why we say that the condition is local. A very convenient and natural condition that implies condition L is that the partial derivatives $\partial f_i/\partial y_j$ exist and are continuous in D.

For the ideas used in the following proof the reader should consult a companion volume on metric spaces.

Proposition 6.1 *Let f satisfy condition L. If a solution to the initial value problem exists on an open interval I containing x_0, then it is the only solution on the interval I.*

Proof Let there be two solutions to the initial value problem, $y = \phi(x)$ and $y = \psi(x)$, on the interval I. Let

$$M = \{x \in I : \phi(x) = \psi(x)\}.$$

We have $x_0 \in M$, by the initial condition that is satisfied by both solutions. Furthermore M is closed relative to I because ϕ and ψ are continuous. But M is also open relative to I. For suppose that $x_1 \in I$, so that $\phi(x_1) = \psi(x_1)$. By the local uniqueness theorem we must have $\phi(x) = \psi(x)$ for all x in an interval containing x_1. Hence M is open. We conclude, since I is connected, that $M = I$. □

Now that we know that the solution is unique as far as it exists we can define a *maximal solution* on an interval $I_{\max} =]\alpha, \omega[$, defined to be the union of all intervals I, that contain x_0, and on which the initial value problem has a solution. On $I_{\max}$ there exists a unique solution, and it cannot be extended to a larger interval. Note that α could be $-\infty$ and ω could be $+\infty$.

It is a fact that neither the maximal forward solution curve (the part with $x_0 < x < \omega$) nor the maximal backward solution curve (the part with $\alpha < x < x_0$) can lie wholly in any compact subset of D.

Exercise Prove the two claims of the last paragraph.

Hint Let the maximal solution be $\phi :]\alpha, \omega[\to \mathbb{R}^n$, and, assuming that the maximal forward solution curve lies inside a compact subset A of D, show that the limit $\zeta := \lim_{x\to\omega-} \phi(x)$ exists. Then consider the solution curve through (ω, ζ) and obtain a contradiction.

This is the conclusion given in many of the books on differential equations and is quite sufficient for most studies of the maximal solution. However, one can obtain a stronger conclusion with a little additional effort. It shows that as x tends to α or ω, the maximal solution curve exits every compact set, without subsequent return. This excludes the possibility that the forward solution curve, whilst not lying wholly inside any compact set, exhibits oscillations as x tends to a finite ω, such that the magnitude of the oscillations increases without bound. We shall need to use the precise information conveyed by Proposition 4.1 about the interval of existence of a solution.

Proposition 6.2 *The maximal solution* $\phi :]\alpha, \omega[\to \mathbb{R}^n$ *has the following property: for every compact* $A \subset D$ *there exist* α' *and* ω' *satisfying* $\alpha < \alpha' < x_0 < \omega' < \omega$, *such that*

$$(x, \phi(x)) \in D \setminus A,$$

for all $x \in]\alpha, \alpha'[$ *and all* $x \in]\omega', \omega[$.

Proof We shall show that ω' exists. A similar argument will deal with the existence of α'. Note that if $\omega = \infty$ the result follows from the fact that compact sets are bounded. We therefore assume that ω is finite.

We proceed by contradiction. Let A be a compact subset of D and assume that no point ω' exists in the interval $]x_0, \omega[$ with the desired properties. It follows that there exists an increasing sequence $(t_n)_{n=1}^{\infty}$ in $]x_0, \omega[$, such that $t_n \to \omega$ and $(t_n, \phi(t_n)) \in A$. Since A is compact, the sequence $\big((t_n, \phi(t_n))\big)_{n=1}^{\infty}$ has a convergent subsequence, so we may assume, without loss of generality, that it is itself convergent. Its limit is of the form (ω, ζ), for some ζ, and the limit is in A.

Choose a closed rectangle[1] S with centre (ω, ζ), such that $S \subset D$. Let $M = \sup_S |f|$ and choose $a > 0$ and $b > 0$, such that $a < b/M$, and such that the rectangle

$$R_\infty := \big\{(x, y) : |x - \omega| \le a,\ |y - \zeta| \le b\big\}$$

is included in the interior of S (we don't want R_∞ to meet the boundary edges or faces of S). If we set

$$R_n := \big\{(x, y) : |x - t_n| \le a,\ |y - \phi(t_n)| \le b\big\}$$

then R_n is a translate of R_∞, and since $t_n \to \omega$ and $\phi(t_n) \to \zeta$, there exists N, such that $R_n \subset S$ for all $n \ge N$. For such n, Picard's theorem (Proposition 4.1) tells us that the initial value problem

$$y' = f(x, y), \quad y(t_n) = \phi(t_n)$$

has a unique solution on the interval $]t_n - a, t_n + a[$, and this solution is therefore identical to ϕ on the intersection of their domains. Note that the length of this interval is independent of n. Since $t_n \to \omega$ the solution of this initial value problem extends $\phi(x)$ to the right of ω when n is sufficiently large. This contradicts the definition of ω. □

The proof is illustrated in Fig. 6.1. In the picture it appears that when $n \ge 5$ the rectangle R_n falls within S, and then the solution curve with initial data $(t_n, \phi(t_n))$

[1] Although we suggestively refer to S as a rectangle, it might be more accurate if $n \ge 2$ to call it a cylinder.

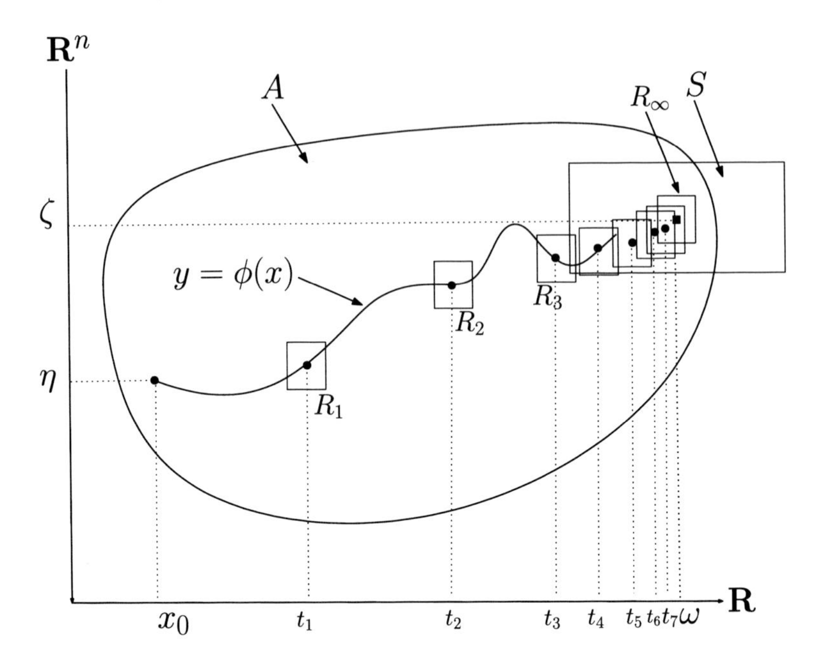

Fig. 6.1 Illustration of the proof

must pass through the vertical sides of R_n. With good eyesight, one can observe that the solution with initial data $(t_7, \phi(t_7))$ extends above $x = \omega$.

One consequence of Proposition 6.2 is an argument or method that can be used to show that an initial value problem has a solution on a preassigned interval $]a, b[$, possibly much larger than that yielded by Picard's theorem. In fact a could be $-\infty$ and b could be $+\infty$. This makes the method particularly useful.

Let us introduce some suggestive geometrical terminology. Given an open set $D \subset \mathbb{R} \times \mathbb{R}^n$ and a function $\phi : I \to \mathbb{R}^n$, where I is an open interval and the graph of ϕ lies in D, we shall say that the curve $y = \phi(x)$ *escapes to the edge of* D if, for every compact $A \subset D$, the curve exits A, without subsequent return, as x tends to the endpoints of I. Thus, for the initial value problem with domain D, Proposition 6.2 says that the maximal solution curve escapes to the edge of D.

Consider then the initial value problem

$$y' = f(x, y), \quad y(x_0) = \eta$$

where f has the domain D. We wish to show that a solution exists on a given interval $]a, b[$, where $a < x_0 < b$. It is important that the cases $a = -\infty$ and $b = +\infty$ are allowed. All we have to do is show that, for each a' and b' that satisfy $a < a' < x_0 < b' < b$, an assumed solution curve on the interval $]a', b'[$ cannot escape to the edge of D. If we can do this it will follow that a solution exists on $]a, b[$. Putting it in the most intuitive fashion: the solution can be extended as long as it cannot escape to the edge of D. Expressed in this way, the principle just enunciated is a simple instance of the method of a priori bounds, that plays an important role in the theory of partial differential equations.

There are some logical difficulties in the principle as formulated above, caused by the fact that a false premise implies any conclusion. Thus if no solution exists on the interval $]a', b'[$, then an assumed solution on this interval both escapes to the edge of D—and does not escape to the edge of D—by default! This is easily remedied by means of further abstraction. For each interval J that contains x_0, let U_J be the set of all curves in D with domain J that escape to the edge of D; and let S_J be the set of solutions of the initial value problem with domain J. The set S_J could be empty. The principle of the previous paragraph can be expressed as follows. If, for all a' and b' such that $a < a' < x_0 < b' < b$, we have $U_J \cap S_J = \emptyset$, where $J =]a', b'[$, then a solution exists on $]a, b[$.

A simple example will illustrate the method and at the same time prove an important result. We are going to consider a linear system in $\mathbb{R}^n$. Let $A(x)$ be an $n \times n$ matrix valued function on the interval $I :=]a, b[$. The interval may be unbounded. The domain D is the set $I \times \mathbb{R}^n$. We suppose the function $A(x)$ to be continuous in I (this can be understood as meaning that the entries are continuous functions, but it is preferable to use a norm in the vector space of matrices to define a metric).

Let $a < x_0 < b$, let $\eta \in \mathbb{R}^n$, and consider the initial value problem

$$y' = A(x)y, \quad y(x_0) = \eta.$$

We claim that a solution exists on $]a, b[$. In fact let $a < a' < x_0 < b' < b$ and let $M = \sup_{a' \le x \le b'} \|A(x)\|$ (using here any convenient matrix norm). Then M is finite since the interval $[a', b']$ is compact. Suppose that a solution $y(x)$ exists on the interval $]a', b'[$. Then, for $a' < x < b'$ we have

$$|y'(x)| \le M|y(x)|$$

and so

$$|y(x)| \le |\eta|\, e^{M|x-x_0|}.$$

Exponential growth may be proverbially fast, but it does not allow the solution curve to escape to infinity on a bounded interval. We conclude that the solution may be extended to the interval $]a, b[$, which, we repeat, may be unbounded.

This result can also be obtained by showing that Picard successive approximations converge on the whole interval $]a, b[$, much as in the discussion of the problem $y' = p(x)y$ given in Sect. 4.1. This is an important result. However, the application of Proposition 6.2 illustrates a typical approach, that of using the differential equation to bound the growth of an assumed solution.

6.1.1 Exercises

1. For the Bernoulli equation $y' - y = xy^2$ (Sect. 1.2 Exercise 3) the general solution can be written as

$$y(x) = \frac{1}{1 - x + Ce^{-x}}.$$

All solutions are of this form except the trivial solution $y = 0$ (in some sense that corresponds to having $C = \infty$). Use this to study the interval of existence of the solution that satisfies the Cauchy condition $y(0) = a$. You should find that for some values of a the solution exists on a half-bounded interval of the form $]-\infty, B[$, for others on a bounded interval of the form $]-A, B[$, whilst for yet other values it exists on $]-\infty, \infty[$. Investigate this. Draw some solution curves showing all the possibilities.

2. In this exercise and the following one we study the Cauchy problem

$$y' = x^2 + y^2, \quad y(0) = 0.$$

The object is to show by very simple estimates that the largest interval to which the solution can be extended is of the form $]-\omega, \omega[$ and derive the bounds $1.253 < \omega < 2.309$.

The largest interval in which the Picard existence theorem guarantees a solution is $[-\frac{1}{\sqrt{2}}, \frac{1}{\sqrt{2}}] \approx [-0.707, 0.707]$ (see Sect. 4.1 Exercise 1). Let us denote the solution by $y = \phi(x)$.

(a) Show that $\phi(x)$ is an odd function, that is $\phi(-x) = -\phi(x)$.
Hint Show that $\psi(x) := -\phi(-x)$ satisfies the same Cauchy problem.

(b) Let $]-\omega, \omega[$ be the largest interval to which the solution can be extended. At this point we do not know yet whether ω is finite. Show that $\phi(x) > x^3/3$ for $0 < x < \omega$.

(c) Let $0 < a < b < \omega$. Show that

$$\frac{1}{\phi(b)} \leq \frac{1}{a^3} + a - b$$

and deduce that if $0 < a < \omega$ then $\omega < a + \dfrac{3}{a^3}$. In particular $\omega < \infty$.
Hint $\phi'(x) > \phi(x)^2$.

(d) Deduce from the previous item that

$$\omega < \min_{a>0} \left(a + \frac{3}{a^3} \right) = \frac{4}{\sqrt{3}} = 2.309\ldots$$

(e) Prove that

$$\lim_{x \to \omega -} \phi(x) = +\infty.$$

(f) Prove that

$$\omega > \sqrt{\frac{\pi}{2}} = 1.253\ldots$$

Hint $\phi'(x) < \omega^2 + \phi(x)^2$ for $0 < x < \omega$.

3. Identify and evaluate the point ω from the previous exercise. Using Sects. 1.2 Exercise 16, and 3.2 Exercise A6, show that the solution of the initial value problem $y' = x^2 + y^2$, $y(0) = 0$, can be expressed in terms of Bessel functions by:

$$y(x) = -\frac{\frac{d}{dx}\left(x^{\frac{1}{2}} J_{-\frac{1}{4}}(x^2/2)\right)}{x^{\frac{1}{2}} J_{-\frac{1}{4}}(x^2/2)}$$

Deduce that $\omega^2/2$ is the lowest positive zero of the Bessel function $J_{-\frac{1}{4}}(x)$. This gives $\omega = 2.003147$ (using a numerical Bessel calculator).

4. The system of equations

$$\frac{dx}{dt} = y, \quad \frac{dy}{dt} = -\omega(t)^2 x$$

describes oscillations of small amplitude of a pendulum, whose length is varying with time. Suppose that $\omega(t)$ is C^1 and bounded. Show that all solutions can be extended to the interval $-\infty < t < \infty$.
Hint Let $r = \sqrt{x^2 + y^2}$ and show that along a solution curve $dr/dt < Kr$ where K is a constant.

5. We revisit Newton's equation (Sect. 2.3)

$$\frac{d^2x}{dt^2} = F(x)$$

where $F(x)$ is a C^1 function defined on the real line $\mathbb{R}$. Let $U(x)$ (potential energy) be an antiderivative of $-F(x)$. Assume that $U(x)$ is bounded below; that is, there exists K, such that $U(x) > K$ for all x. Prove that every solution can be extended to the interval $-\infty < t < \infty$.
Hint Show that $|dx/dt|$ satisfies a bound, independent of t.

6.2 Dependence on Initial Conditions

The solution of the initial value problem is not just a function of x; it is also a function of the initial value. The interval of definition of the maximal solution also depends on the initial value. Suppose that the initial value problem,

$$y' = f(x, y), \quad y(x_0) = \eta$$

has, for each η in a given set A, a solution on a common interval $]a, b[$ (allowing, as is usual, the cases $a = -\infty$ or $b = \infty$). We can write the solution as a function $\phi_{x_0}^{x}(\eta)$. That is

$$\frac{d}{dx}\phi_{x_0}^{x}(\eta) = f(x, \phi_{x_0}^{x}(\eta)), \quad \phi_{x_0}^{x_0}(\eta) = \eta, \quad (a < x < b, \ \eta \in A).$$

The notation $\phi_{x_0}^{x}(\eta)$ was introduced in Sect. 5.3 for linear equations with variable coefficients. It suggests the evolution of a system with a passage of time from x_0 to x, the initial state being η and the final state $\phi_{x_0}^{x}(\eta)$. In Sect. 5.3 we could define a linear mapping $\phi_{x_0}^{x}$. The situation is now more complicated, but we can study the family of mappings $\phi_{x_0}^{x} : A \to \mathbb{R}$, indexed by $x \in\ ]a, b[$. Intuitively these mappings advance time from x_0 to x (though $x < x_0$ is allowed). We shall show that these mappings are continuous, indeed Lipschitz continuous, and that they are C^1 if f is C^1. These conclusions have practical significance.

One may go further and consider $\phi_{x_0}^{x}(\eta)$ as a function of the triplet (x_0, x, η). We leave it to the reader to extend the results to these cases, while, for the most part, limiting ourselves here to varying η alone.

Proposition 6.3 *Suppose that $f : D \to \mathbb{R}$ satisfies condition L. Suppose further that the initial value problem*

$$y' = f(x, y), \quad y(x_0) = \eta$$

has a solution on the interval $]a, b[$. Let $a < a' < x_0 < b' < b$. Then there exist $\delta > 0$ and $M > 0$, such that for $|h| < \delta$, the initial value problem

$$y' = f(x, y), \quad y(x_0) = \eta + h$$

has a solution on the interval $]a', b'[$, and for $a' < x < b'$ we have

$$|\phi_{x_0}^{x}(\eta + h) - \phi_{x_0}^{x}(\eta)| \le M|h|.$$

Note that a could be $-\infty$, b could be $+\infty$, but a' and b' are finite numbers. The proof is illustrated in Fig. 6.2.

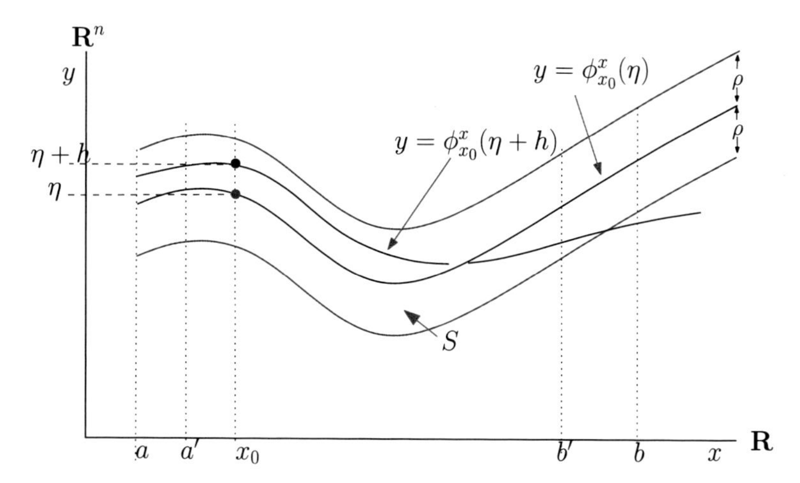

Fig. 6.2 Illustration of the proof

Proof Let $\rho > 0$ be chosen so that the set

$$S := \{(x, y) : |y - \phi_{x_0}^x(\eta)| \leq \rho,\ a' \leq x \leq b'\}$$

is included in the domain D of f. Let K be a Lipschitz constant for f in the domain S, as provided for by condition L. Let $|h| < \rho$. The solution $\phi_{x_0}^x(\eta + h)$, which passes through $(x_0, \eta + h)$, satisfies

$$\phi_{x_0}^x(\eta + h) = \eta + h + \int_{x_0}^x f\big(s, \phi_{x_0}^s(\eta + h)\big)\, ds$$

as long as it can be extended. Therefore,

$$\phi_{x_0}^x(\eta + h) - \phi_{x_0}^x(\eta) = h + \int_{x_0}^x \big(f(s, \phi_{x_0}^s(\eta + h)) - f(s, \phi_{x_0}^s(\eta))\big)\, ds$$

and so, for $x_0 < x < b'$, and provided the solution curve through $(x_0, \eta + h)$ remains within S, we have

$$\begin{aligned} \left|\phi_{x_0}^x(\eta + h) - \phi_{x_0}^x(\eta)\right| &\leq |h| + \int_{x_0}^x \left|f(s, \phi_{x_0}^s(\eta + h)) - f(s, \phi_{x_0}^s(\eta))\right| ds \\ &\leq |h| + K \int_{x_0}^x \left|\phi_{x_0}^s(\eta + h) - \phi_{x_0}^s(\eta)\right| ds. \end{aligned}$$

Hence, by Gronwall's inequality (Lemma 4.3)

$$\left|\phi^x_{x_0}(\eta+h)-\phi^x_{x_0}(\eta)\right| \le |h|e^{K(x-x_0)} < |h|e^{K(b'-x_0)}$$

for $x_0 < x < b'$, as long as the solution curve through $(x_0, \eta+h)$ remains within S.

Let $\delta = \rho e^{-K(b'-x_0)}$. If $|h| < \delta$ the solution curve through $(x_0, \eta+h)$ cannot escape from the domain S on the interval $]x_0, b'[$, provided it can be extended so far; we can conclude, by the fundamental extension principle, that it can indeed be extended so far. Finally, we set $M = e^{K(b'-x_0)}$. □

Lipschitz conditions have figured prominently in this chapter, as also in Chap. 4. Now we need to consider functions that are *locally Lipschitz continuous*. The property is stronger than continuity, but weaker than differentiability. As it may be unfamiliar to the reader we give a formal definition.

Definition A mapping $f : A \to \mathbb{R}^n$, with domain $A \subset \mathbb{R}^n$, is said to be *locally Lipschitz continuous* if, for each y in the domain of f, there exists $\delta > 0$ and $M \ge 0$, such that for all $y^{(1)}$ and $y^{(2)}$ in A that satisfy $|y^{(1)} - y| < \delta$ and $|y^{(2)} - y| < \delta$ we have $|f(y^{(1)}) - f(y^{(2)})| \le M|y^{(1)} - y^{(2)}|$.

It may not be possible to specify a single Lipschitz constant that works throughout A. The reader may like show that for each compact $B \subset A$ there exists a Lipschitz constant that works for B.

Proposition 6.4 *Suppose that $f : D \to \mathbb{R}$ satisfies condition L. Let A be a subset of $\mathbb{R}^n$ and suppose that for each η in A the initial value problem*

$$y' = f(x, y), \quad y(x_0) = \eta$$

has a solution on the interval $]a, b[$. Then the mappings $\phi^x_{x_0} : A \to \mathbb{R}$, indexed by $x \in]a, b[$, are all locally Lipschitz continuous.

Proof We repeat the calculations from the proof of Proposition 6.3, but extract more information. Let $\eta \in A$ and let a' and b' be such that $a < a' < x_0 < b' < b$. Choose ρ, S, K and δ as in the proof of Proposition 6.3. Let h and k satisfy $|h| < \delta$ and $|k| < \delta$ and consider the solutions through $(x_0, \eta+h)$ and $(x_0, \eta+k)$. For all x in the interval $x_0 < x < b'$, such that these solutions can be extended up to x, we have

$$\left|\phi^x_{x_0}(\eta+h)-\phi^x_{x_0}(\eta+k)\right| \le |h-k| + K\int_{x_0}^{x}\left|\phi^s_{x_0}(\eta+h)-\phi^s_{x_0}(\eta+k)\right| ds$$

By Gronwall's inequality (Lemma 4.3) we have

$$\left|\phi^x_{x_0}(\eta+h)-\phi^x_{x_0}(\eta+k)\right| \le |h-k|e^{K(x-x_0)} < |h-k|e^{K(b'-x_0)}.$$

The conclusion follows for $x > x_0$. The case $x < x_0$ is left to the reader. □

Recall from analysis that given a metric space M a function f, with domain M and codomain $\mathbb{R} \cup \{-\infty\} \cup \{+\infty\}$, is said to be upper semi-continuous if the inverse image $f^{-1}\big([-\infty, a[\big)$ is open for every a, whilst it is said to be lower semi-continuous if $f^{-1}\big(]a, +\infty]\big)$ is open for every a.

Proposition 6.5 *Suppose that $f : D \to \mathbb{R}$ satisfies condition L. Then the endpoints α and ω of the interval of definition of the maximal solution of the initial value problem*

$$y' = f(x, y), \quad y(x_0) = \eta$$

are respectively upper semi-continuous, and lower semi-continuous, functions of η.

Proof Let the maximal solution curve through (x_0, η) be defined on the interval $]\alpha(\eta), \omega(\eta)[$. By Proposition 6.3, for each real β, the set of all η such that $\omega(\eta) > \beta$ is open, as is also the set of all η such that $\alpha(\eta) < \beta$. □

With little additional effort we can obtain continuity of the solution as a function of the pair (x, η). This will be needed at a decisive point in Chap. 7.

Proposition 6.6 *Suppose that $f : D \to \mathbb{R}$ satisfies condition L. Suppose that the solution of the initial value problem*

$$y' = f(x, y), \quad y(x_0) = \eta$$

can be extended beyond a point $b > x_0$. Let $\varepsilon > 0$. Then there exists $\delta > 0$, such that if $|h| < \delta$, the solution of the initial value problem

$$y' = f(x, y), \quad y(x_0) = \eta + h$$

can be extended beyond $b + \delta$, and satisfies

$$|\phi_{x_0}^{b'}(\eta + h) - \phi_{x_0}^{b}(\eta)| \leq \varepsilon$$

for $|b - b'| < \delta$.

Proof Study Fig. 6.3. Let $y(x)$ be the solution through (x_0, η). Construct the "cylinder" C with "base" $x = b - \delta_1$, "top" $x = b + \delta_1$, and "mantle" $|y - y(b)| = \varepsilon$, choosing δ_1 sufficiently small that the solution curve $y = y(x)$ can be extended beyond $b + \delta_1$, and such that it enters the cylinder via its base and departs via its top. Now construct the tube S as described in the proof of Proposition 6.3, choosing ρ sufficiently small that S enters and leaves the cylinder C via its base and its top. Choose δ_2 so that if $|h| < \delta_2$ the solution through $(x_0, \eta + h)$ remains in the tube S beyond $x = b + \delta_1$. Finally let $\delta = \min(\delta_1, \delta_2)$. □

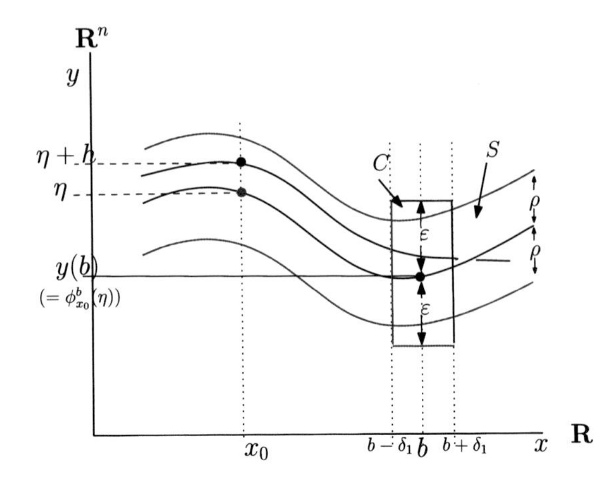

Fig. 6.3 The proof of Proposition 6.6

We shall describe some examples illustrating the ideas of this section. For the most part they are strikingly simple, in contrast to the rather opaque proofs just given.

Example Consider the problem $y' = y$, for $y \in \mathbb{R}$. We have $\phi_0^x(\eta) = e^x\eta$, for $\eta \in \mathbb{R}$. The continuity of ϕ_0^x is obvious, but there is an interesting point. For fixed η and h the difference $\phi_0^x(\eta + h) - \phi_0^x(\eta) = e^x h$ grows exponentially with increasing x. Solution curves pull rapidly apart from each other and from a practical viewpoint the continuity of ϕ_0^x can become rather meaningless for large x. Moreover, in practical cases, it may be more important that the "final destination," that is, the limit $\lim_{x\to\infty} \phi_0^x(\eta)$, is a discontinuous function of η.

Example/Exercise For the problem $y' = 2xy^2$ derive the formula

$$\phi_0^x(\eta) = \frac{\eta}{1 - \eta x^2}.$$

Show that the right-hand endpoint ω of the maximal interval of definition of the solution $\phi_0^x(\eta)$ is given by

$$\omega = \begin{cases} \infty, & (\eta \le 0) \\ 1/\sqrt{\eta}, & (\eta > 0) \end{cases}$$

In this example (and in the previous one) ω is a continuous function of η with values in the extended real line.

Example/Exercise The problem $y' = y + xy^2$ was studied in Sect. 6.1 Exercise 1. Show that

$$\phi_0^x(\eta) = \frac{\eta}{\eta(1-x) + (1-\eta)e^{-x}}$$

Show that for each η, the left and right endpoints of the maximal interval, α and ω, are respectively the negative and positive values of ξ, if any, that satisfy

$$(\xi - 1)e^{\xi} = \frac{1}{\eta} - 1$$

with $\alpha = -\infty$ if no negative value exists, and $\omega = \infty$ if no positive value exists.

Exercise The same equation exhibits an instance when an endpoint of the maximal interval is a discontinuous function of the initial value. If you have drawn sufficiently accurate pictures in Sect. 6.1 Exercise 1 you may be able to see this. Start by computing the solution $\phi_1^x(\eta)$ (the one with initial value η at $x = 1$). For $\eta = -e/k$, with $k > 0$, show that this is the solution curve

$$y = \frac{1}{1 - x - ke^{-x}}$$

Show that the maximum interval is $]0, \infty[$ if $k = 1$, but $]-\infty, \infty[$ if $k > 1$. Hence, for $x_0 = 1$, α is a discontinuous function of η at $\eta = -e$.

6.2.1 *Differentiability of* $\phi_{x_0}^x$

The proposition asserting that the solution of the initial value problem is a differentiable function of the initial value, when $f(x, y)$ is a differentiable function of y, can appear one of the hardest in the "essential" theory of differential equations. This subsection and the next are possibly the most difficult reading in this text, but the results are immensely important, both theoretically and practically, as we shall see.

The arguments in the coming Proposition 6.7 could be written out for partial derivatives, as opposed to total derivatives, which are matrices. However, as we have been much occupied in previous chapters with matrix differential equations, it is very natural to use them here. We recall from multivariate analysis that the (total) derivative of a mapping $f : \mathbb{R}^n \to \mathbb{R}^n$ at a point y, if it exists, is the unique linear mapping $Df(y)$ (or matrix) with the following property:

For all $\varepsilon > 0$ there exists $\delta > 0$, such that for all vectors h that satisfy $|h| < \delta$ we have

$$\big|f(y+h) - f(y) - Df(y)h\big| \le \varepsilon|h|.$$

We also recall that if f is differentiable at the point y, then all partial derivatives $\partial f_i/\partial y_j$ exist at y and the matrix of $Df(y)$ is the Jacobian matrix

$$\left[\frac{\partial f_i}{\partial y_j}\right]_{j=1\ i=1}^{n\quad n}.$$

Conversely, if the partial derivatives exist in an open set A and are continuous in A, then f is differentiable in A. In this important and practical case we say that f is continuously differentiable, or C^1, in A.

If we consider the derivative at each of a set of points A, instead of a single point y, then the value of δ for a given ε may depend on the point in question. It may not be possible to find one value of δ that works simultaneously for the given ε at all points in A. However, if f is a C^1 function, then uniformity can be achieved for compact subsets of A.

In the context of the differential equation $y' = f(x, y)$, we shall need to apply these ideas to the derivative $D_y f$, the derivative of the mapping $y \mapsto f(x, y)$, holding x fixed. As usual f is defined in an open subset D of $\mathbb{R} \times \mathbb{R}^n$, where x ranges over the first factor and y over the second. A precise statement of what we need in this context is as follows:

Suppose that the partial derivatives of f with respect to the coordinates of y are continuous in the domain $D \subset \mathbb{R} \times \mathbb{R}^n$ and let S be a compact subset of D. Then for all $\varepsilon > 0$ there exists $\delta > 0$, such that for all vectors $h \in \mathbb{R}^n$ that satisfy $|h| < \delta$ and for all $(x, y) \in S$ for which the segment joining (x, y) and $(x, y + h)$ lies in S, we have

$$\big|f(x, y+h) - f(x, y) - D_y f(x, y)h\big| \le \varepsilon|h|.$$

The reader with good knowledge of multivariate calculus should have no difficulty proving it, using the mean value theorem and the boundedness of continuous functions on compact sets. It will be used at key points in the proofs of several subsequent propositions usually in the following context. A solution to the initial value problem is known on the interval $]a, b[$. Given a' and b' that satisfy $a < a' < x_0 < b' < b$, and $\rho > 0$, we construct the set

$$S := \left\{(x, y) : |y - \phi_{x_0}^x(\eta)| \le \rho,\ a' \le x \le b'\right\}$$

(as in the proof of Proposition 6.3). If ρ is sufficiently small, S will lie inside D (the domain of f) and we can apply the previous paragraph to S. Usually we will not mention S explicitly, but merely require $|h|$ to be "sufficiently small".

Proposition 6.7 *Suppose that $f : D \to \mathbb{R}^n$ is continuous and that the partial derivatives of f with respect to the coordinates of y are continuous in the domain $D \subset \mathbb{R} \times \mathbb{R}^n$. Let A be an open subset of $\mathbb{R}^n$ and suppose that for each η in A the initial value problem*

$$y' = f(x, y), \quad y(x_0) = \eta$$

has a solution on the interval $]a, b[$. Then:

1. *The mappings $\phi^x_{x_0} : A \to \mathbb{R}^n$, indexed by $x \in\,]a, b[$, are C^1 functions.*
2. *The derivative $T(x) := D\phi^x_{x_0}(\eta)$ (the derivative being taken with respect to η alone) is the solution of the linear initial value matrix problem:*

$$\frac{dT}{dx} = D_y f\big(x, \phi^x_{x_0}(\eta)\big)\, T, \quad T(x_0) = I. \tag{6.1}$$

Proof Equation (6.1), being linear, can be solved on the interval $]a, b[$. Its solution is a matrix function, written here simply as $T(x)$, (we will not vary x_0 or η so it is unnecessary to keep track of them). The first, and longer, part of the proof is the verification that $T(x) = D\phi^x_{x_0}(\eta)$.

Let $x_0 < c < b$ and let h be a vector, in length sufficiently small that the solution curve through $(x_0, \eta + h)$ can be extended through c. Referring to Proposition 6.2 for the details, we can supply a constant M, such that for $x_0 \le x \le c$ and all sufficiently small vectors h, we have the Lipschitz estimate

$$|\phi^x_{x_0}(\eta + h) - \phi^x_{x_0}(\eta)| \le M|h|.$$

Then, for all x in the interval $x_0 \le x \le c$ we have:

$$\frac{d}{dx}\Big(\phi^x_{x_0}(\eta + h) - \phi^x_{x_0}(\eta) - T(x)h\Big) = f\big(x, \phi^x_{x_0}(\eta + h)\big) - f\big(x, \phi^x_{x_0}(\eta)\big) - D_y f\big(x, \phi^x_{x_0}(\eta)\big)T(x)h$$

$$\begin{aligned} &= f\big(x, \phi^x_{x_0}(\eta + h)\big) - f\big(x, \phi^x_{x_0}(\eta)\big) \\ &\quad - D_y f\big(x, \phi^x_{x_0}(\eta)\big)\big(\phi^x_{x_0}(\eta + h) - \phi^x_{x_0}(\eta)\big) \\ &\quad + D_y f\big(x, \phi^x_{x_0}(\eta)\big)\big(\phi^x_{x_0}(\eta + h) - \phi^x_{x_0}(\eta) - T(x)h\big) \end{aligned} \tag{6.2}$$

Let $\varepsilon > 0$. Because $D_y f$ is continuous, and by the local Lipschitz continuity of $\phi^x_{x_0}$ (Proposition 6.4), we can choose $\delta > 0$ so small that, given $x_0 \le x \le c$ and $|h| < \delta$, we have

$$\left|f\left(x, \phi_{x_0}^x(\eta+h)\right)-f\left(x, \phi_{x_0}^x(\eta)\right)-D_y f\left(x, \phi_{x_0}^x(\eta)\right)\left(\phi_{x_0}^x(\eta+h)-\phi_{x_0}^x(\eta)\right)\right|$$
$$\leq \varepsilon\left|\phi_{x_0}^x(\eta+h)-\phi_{x_0}^x(\eta)\right| \leq \varepsilon M|h|. \tag{6.3}$$

(We refer the reader to the paragraph preceeding the proposition for more detailed explanation.) Let L be a bound for the matrix norm of $D_y f(x, \phi_{x_0}^x(\eta))$ in the interval $x_0 \leq x \leq c$. We combine (6.2) and (6.3) to obtain

$$\left|\frac{d}{dx}\left(\phi_{x_0}^x(\eta+h)-\phi_{x_0}^x(\eta)-T(x)h\right)\right|$$
$$\leq \varepsilon M|h|+L\left|\phi_{x_0}^x(\eta+h)-\phi_{x_0}^x(\eta)-T(x)h\right|$$

Using Gronwall's inequality (Lemma 4.3) we now obtain, for $|h|<\delta$ and $x_0 \leq x \leq c$:

$$\left|\phi_{x_0}^x(\eta+h)-\phi_{x_0}^x(\eta)-T(x)h\right| \leq \varepsilon M e^{L(x-x_0)}|h|.$$

This says that $T(x)$ is the derivative of $\phi_{x_0}^x$ at η.

It remains to show that $D\phi_{x_0}^x(\eta)$ is a continuous function of η. Let $x_0<c<b$ and let the vector h be sufficiently small that the solution through $(x_0, \eta+h)$ can be extended beyond $x=c$. Then

$$\frac{d}{dx} D\phi_{x_0}^x(\eta)=D_y f\left(x, \phi_{x_0}^x(\eta)\right) D\phi_{x_0}^x(\eta)$$

and

$$\frac{d}{dx} D\phi_{x_0}^x(\eta+h)=D_y f\left(x, \phi_{x_0}^x(\eta+h)\right) D\phi_{x_0}^x(\eta+h)$$

whence

$$\frac{d}{dx}\left(D\phi_{x_0}^x(\eta+h)-D\phi_{x_0}^x(\eta)\right)$$
$$=\left(D_y f\left(x, \phi_{x_0}^x(\eta+h)\right)-D_y f\left(x, \phi_{x_0}^x(\eta)\right)\right) D\phi_{x_0}^x(\eta)$$
$$+D_y f\left(x, \phi_{x_0}^x(\eta+h)\right)\left(D\phi_{x_0}^x(\eta+h)-D\phi_{x_0}^x(\eta)\right)$$

Let $\varepsilon>0$. Since $D_y f$ is continuous, and hence uniformly continuous on compact sets, there exists $\delta>0$ such that, for $|h|<\delta$ the first summand on the right-hand side is less than ε for all x in the interval $[x_0, c]$. Letting M be a bound for

$D_y f(x, \phi_{x_0}^x(\eta + h))$ valid for $x_0 \leq x \leq c$ and $|h| < \delta$, we deduce that

$$\left|\frac{d}{dx}\Big(D\phi_{x_0}^x(\eta + h) - D\phi_{x_0}^x(\eta)\Big)\right| \leq \varepsilon + M\left|D\phi_{x_0}^x(\eta + h) - D\phi_{x_0}^x(\eta)\right|.$$

Now Gronwall's inequality implies

$$\left|D\phi_{x_0}^x(\eta + h) - D\phi_{x_0}^x(\eta)\right| \leq \varepsilon e^{M(x-x_0)}$$

provided $|h| < \delta$, thus exhibiting the continuity of $D\phi_{x_0}^x(\eta)$. □

The linear differential equation (6.1) satisfied by $D\phi_{x_0}^x(\eta)$ is called the *equation of variation*. An equivalent version says that, for $k = 1, \dots, n$, the partial derivative $p(x) := (\partial/\partial\eta_k)\phi_{x_0}^x(\eta)$ satisfies the initial value problem

$$\frac{dp}{dx} = D_y f\big(x, \phi_{x_0}^x(\eta)\big)\, p, \quad p(x_0) = e_k \tag{6.4}$$

where e_k denotes the kth basis vector of $\mathbb{R}^n$. It is natural to write this as

$$\frac{dp}{dx} = D_y f(x, y)\, p, \quad p(x_0) = e_k \tag{6.5}$$

where y is understood to be the solution of $y' = f(x, y)$ with initial value η at $x = x_0$. We shall see that this equation can be imbedded into a potentially infinite array of equations that determine the higher derivatives of $\phi_{x_0}^x(\eta)$ with respect to η, given that $f(x, y)$ has derivatives of sufficiently high order with respect to y, that are continuous in the domain D.

Exercise For the problem $y' = xy^2$ with initial condition $y = \eta$ at $x = 0$, show that the equation of variation is

$$p' = 2xyp, \quad p(0) = 1$$

where p is the derivative of the solution with respect to η, and deduce that

$$p = \frac{y^2}{\eta^2}.$$

This formula is shorthand (of a kind frequently used, and which one should get accustomed to) for

$$\frac{\partial}{\partial\eta}\phi_0^x(\eta) = \frac{(\phi_0^x(\eta))^2}{\eta^2}.$$

6.2.2 Higher Derivatives of $\phi^x_{x_0}$

The pair $\big(\phi^x_{x_0}(\eta), D\phi^x_{x_0}(\eta)\big)$ is the solution to the initial value problem

$$\begin{aligned} y' &= f(x, y), & y(x_0) &= \eta \\ T' &= D_y f(x, y)\, T, & T(x_0) &= I \end{aligned} \tag{6.6}$$

uniting the original initial value problem and the equation of variation into one system (compare (6.5)). The dependent variable is the pair (y, T) consisting of a vector and a matrix; it lives in a space of dimension $n + n^2$. Suppose that f has second order partial derivatives with respect to the coordinates of y, and they are continuous in D. Then the mapping

$$(y, T) \mapsto (f(x, y), D_y f(x, y)T),$$

that defines the direction field of (6.6), is a C^1 function of (y, T), with derivatives that are continuous functions of (x, y, T). We can apply Proposition 6.7 to (6.6) and deduce that *its* solution is a C^1 function of η. In particular $\phi^x_{x_0}$ is a C^2 function.

It seems obvious that this can be made the basis of an induction argument, that should lead to the conclusion:

If f has continuous partial derivatives with respect to the coordinates of y up to order m, then $\phi^x_{x_0}$ is a C^m mapping.

This implies:

If f has continuous partial derivatives with respect to the coordinates of y of all orders, then $\phi^x_{x_0}$ is a C^∞ mapping.

Making the induction precise is a little tricky. It seems to require us to write down the analogues of (6.6) for higher derivatives. We would need to compute the nth order partial derivatives

$$\frac{\partial^\alpha}{\partial \eta^\alpha} f(x, \phi^x_{x_0}(\eta))$$

with $|\alpha| = n$. Here we are using *multi-index notation*, in which $\alpha = (\alpha_1, \ldots, \alpha_n)$ is a vector whose coordinates are non-negative integers. We define the differential operator

$$\frac{\partial^\alpha}{\partial \eta^\alpha} = \left(\frac{\partial}{\partial \eta_1}\right)^{\alpha_1} \cdots \left(\frac{\partial}{\partial \eta_n}\right)^{\alpha_n}.$$

The order of a multi-index is defined as $|\alpha| := \alpha_1 + \cdots + \alpha_n$ (the Euclidean length is not meant here in spite of the notation).

The notation of multi-indices is convenient for induction proofs because

$$\frac{\partial^\alpha}{\partial\eta^\alpha}\frac{\partial^\beta}{\partial\eta^\beta} = \frac{\partial^{\alpha+\beta}}{\partial\eta^{\alpha+\beta}}$$

but can appear odd in simple calculations; for example what we would normally write as $\partial/\partial\eta_1$ is written using multi-indices as $\partial^{e_1}/\partial\eta^{e_1}$, where e_1 is the basis vector $(1, 0, \ldots, 0)$.

There exist complicated formulas for the nth order derivatives of a composed mapping, associated with the name Faà di Bruno. However, all we need here is some general idea of its structure, to enable us to apply Proposition 6.7 inductively. The reader may be able to convince themselves that, provided the derivatives exist to the required orders, and if $|\alpha| = m$, then each component of the n-vector

$$\frac{\partial^\alpha}{\partial\eta^\alpha} f(x, \phi^x_{x_0}(\eta))$$

(remember: it's an n-vector because f takes values in $\mathbb{R}^n$) can be expressed as a polynomial function in the components of the partial derivatives

$$\frac{\partial^\beta}{\partial\eta^\beta}\phi^x_{x_0}(\eta), \quad (1 \leq |\beta| \leq m)$$

and in the components of the partial derivatives of f with respect to the coordinates of y, up to order m, evaluated at $(x, \phi^x_{x_0}(\eta))$, that is to say the derivatives

$$\frac{\partial^\beta f_k}{\partial y^\beta}(x, \phi^x_{x_0}(\eta)), \quad (1 \leq k \leq m,\ 1 \leq |\beta| \leq m).$$

It should also be obvious, though it is not needed for the induction, that the expression is linear in the latter derivatives.

Now we can get an idea of what the m-order equation of variation looks like. If $f(x, y)$ is a C^m function of y, and the solution $\phi^x_{x_0}(\eta)$ is a C^r function of η for some $r < m$, then the vector quantities

$$p_0(x, \eta) := \phi^x_{x_0}(\eta), \quad p_\alpha(x, \eta) := \frac{\partial^\alpha}{\partial\eta^\alpha}\phi^x_{x_0}(\eta), \quad (1 \leq |\alpha| \leq r)$$

satisfy an initial value problem, forming a vast array if r is big, and obtained simply by repeatedly differentiating the equation

$$\frac{d}{dx}\phi^x_{x_0}(\eta) = f(x, \phi^x_{x_0}(\eta))$$

with respect to the coordinates of η, and applying the chain rule. In terms of the vector quantities p_α defined above, this array has the form

$$\frac{dp_0}{dx} = f(x, p_0), \quad \frac{dp_\alpha}{dx} = E_\alpha\big(p_0, (p_\beta)_{1\le|\beta|\le|\alpha|}\big),$$

with initial conditions

$$p_0(0) = \eta, \quad p_{e_k}(0) = e_k \text{ (for } 1 \le k \le n), \quad p_\alpha(0) = 0 \text{ (for } 2 \le |\alpha| \le r),$$

where E_α is a C^{m-r} function of p_0 and a polynomial function in the quantities $(p_\beta)_{1\le|\beta|\le|\alpha|}$. Now the following proposition can be proved by induction on the integer m; we omit the details of the proof, which exploits in a fairly obvious way the discussion of this and the preceding paragraphs.

Proposition 6.8 *Suppose that $f : D \to \mathbb{R}^n$ is continuous and that f has continuous partial derivatives with respect to the coordinates of y up to order m, where $m \ge 1$. Let A be an open subset of $\mathbb{R}^n$ and suppose that for each η in A the initial value problem,*

$$y' = f(x, y), \quad y(x_0) = \eta$$

has a solution $y(x) = \phi^x_{x_0}(\eta)$ on the interval $]a, b[$. Then $\phi^x_{x_0}(\eta)$ is a C^m function of η.

For the problem $y' = f(x, y)$, $y(x_0) = \eta$, in the case that y is a scalar, one can, and should, avoid multi-indices. Set

$$p_k(x) := \frac{\partial^k}{\partial \eta^k}\phi^x_{x_0}(\eta).$$

Then the quantities $p_k(x)$, $k = 0, 1, 2, \ldots$ satisfy a potentially infinite system. It begins

$$\begin{aligned}
p_0' &= f(x, p_0), \quad p_0(x_0) = \eta \\
p_1' &= \frac{\partial f}{\partial y}(x, p_0)p_1, \quad p_1(x_0) = 1, \\
p_2' &= \frac{\partial f}{\partial y}(x, p_0)p_2 + \frac{\partial^2 f}{\partial y^2}(x, p_0)p_1^2, \quad p_2(x_0) = 0 \\
&\vdots \qquad\qquad\qquad\qquad\qquad \vdots
\end{aligned}$$

Exercise What are the next two lines?

In practical mathematics one may want to develop the solution of an initial value problem as a Maclaurin or Taylor series[2] in powers of the initial value, a procedure that is justified by Proposition 6.8. Because second order scalar equations are so frequently encountered in this context, we shall look at an example.

Example Newton's equation with potential $U(x) = \frac{1}{3}x^3 + \frac{1}{2}x^2$ exhibits oscillations near its equilibrium point $x = 0$ (see Sect. 2.3). Consider the problem (with dependent variable x and independent variable t as in Sect. 2.3):

$$\frac{d^2x}{dt^2} + x + x^2 = 0, \quad x(0) = \alpha, \ \frac{dx}{dt}(0) = 0.$$

The solution $x(t, \alpha)$ can be developed as a series

$$x(t, \alpha) = x_0(t) + x_1(t)\alpha + x_2(t)\alpha^2 + \cdots$$

We wish to calculate some of the coefficient functions. To begin with it is more convenient to use dashes to denote differentiation with respect to t. Moreover, we are going to proceed directly with the second order equation, and avoid writing out the equivalent first order plane system. Slightly changing the notation we write the problem as:

$$x'' + x + x^2 = 0, \quad x|_{t=0} = \alpha, \ x'|_{t=0} = 0.$$

We must think of x here as $x(t, \alpha)$. Putting $\alpha = 0$ we at once find $x_0(t) = 0$. Differentiating with respect to α (this and successive differentiations are justified by Proposition 6.8), we find:

$$\left(\frac{\partial x}{\partial \alpha}\right)'' + \frac{\partial x}{\partial \alpha} + 2x\frac{\partial x}{\partial \alpha} = 0, \quad \left.\frac{\partial x}{\partial \alpha}\right|_{t=0} = 1, \ \left.\left(\frac{\partial x}{\partial \alpha}\right)'\right|_{t=0} = 0\,.$$

Putting $\alpha = 0$, which replaces the factor x in the third term by $x_0(t)$, that is by 0, we obtain an initial value problem for $x_1(t)$:

$$x_1'' + x_1 = 0, \quad x_1(0) = 1, \ x_1'(0) = 0$$

so that

$$x_1(t) = \cos t.$$

[2] The use of these terms instead of "power series" avoids any implication of convergence. Even a divergent Taylor series represents its function asymptotically; and in practical mathematics this may be enough.

A second differentiation with respect to α produces:

$$\left(\frac{\partial^2 x}{\partial \alpha^2}\right)'' + \frac{\partial^2 x}{\partial \alpha^2} + 2x\left(\frac{\partial^2 x}{\partial \alpha^2}\right) + 2\left(\frac{\partial x}{\partial \alpha}\right)^2 = 0, \quad \left.\frac{\partial^2 x}{\partial \alpha^2}\right|_{t=0} = 0, \quad \left.\left(\frac{\partial^2 x}{\partial \alpha^2}\right)'\right|_{t=0} = 0$$

Notice how the initial conditions are now homogeneous. In fact the inhomogeneity passes to the differential equation itself, for, putting $\alpha = 0$, and using the fact that $x_1(t) = \cos t$, we obtain an initial value problem for $x_2(t)$:

$$x_2'' + x_2 = -\cos^2 t, \quad x_2(0) = 0, \ x_2'(0) = 0.$$

This gives the solution

$$x_2(t) = -\frac{1}{2} + \frac{1}{3}\cos t + \frac{1}{6}\cos 2t$$

as the reader should check. A further differentiation with to respect to α produces an initial value problem for $x_3(t)$, and so on. It is fairly obvious that $x_k(t)$ (for $k \geq 2$) will satisfy an initial value problem of the form

$$x_k'' + x_k = f_k(t), \quad x_k(0) = x_k'(0) = 0,$$

where $f_k(t)$ is an exponential polynomial, which can be solved by the methods of Sect. 1.4.

An important extension of the material of this section is to equations in which the variable y is a complex vector, that is, $y = (y_1, \ldots, y_n) \in \mathbb{C}^n$. In this situation we shall still treat the independent variable x as real. Since $\mathbb{C}^n$ can be seen as $\mathbb{R}^{2n}$, there is really nothing new involved here, unless $f(x, y)$ is a *complex analytic* function of y. This means that the complex partial derivatives $\partial f/\partial y_k$ exist, with respect to the *complex coordinates* of y.

Proposition 6.9 *Let $D \subset \mathbb{R} \times \mathbb{C}^n$. Suppose that $f : D \to \mathbb{C}^n$ is continuous, and that $f(x, y)$ is a complex analytic function of $y \in \mathbb{C}^n$. Let A be an open subset of $\mathbb{C}^n$ and suppose that for each η in A the initial value problem*

$$y' = f(x, y), \quad y(x_0) = \eta$$

has a solution on the interval $]a, b[$. Then the mappings $\phi_{x_0}^x : A \to \mathbb{C}^n$, indexed by $x \in]a, b[$, are complex analytic functions.

The proof of this proposition uses the same arguments as the proof of Proposition 6.7, replacing the real derivative by the complex derivative. Complex analytic functions have some very strong properties. For example they can be expanded locally in power series in the complex coordinates, so that, in particular they have complex derivatives of all orders. This is probably familiar to the reader in the case of one complex variable.

6.2.3 Equations with Parameters

A problem of the form

$$y' = f(\lambda, x, y), \quad y(x_0) = \eta, \quad (\lambda \in \mathbb{R})$$

in which the direction field depends on a parameter λ, can be reduced to the problem already studied. It leads to an important application of this material to practical mathematics.

We introduce a new dependent variable, namely the pair (λ, y), and consider the system

$$\begin{aligned} \lambda' &= 0, & y' &= f(\lambda, x, y), \\ \lambda(x_0) &= \lambda_0, & y(x_0) &= \eta. \end{aligned}$$

The solution of this gives the solution of the first problem at $\lambda = \lambda_0$.

As a result we deduce, for example, that if $f(\lambda, x, y)$ is a C^∞ function of the pair (λ, y), such that all derivatives are continuous functions of (λ, x, y), then the solution of the initial value problem is a C^∞ function of λ. This has practical consequences, as we shall now see.

A common situation is that we know a solution $y^{(0)}(x)$ of an initial value problem

$$y' = f(x, y), \quad y(x_0) = \eta$$

and we wish to consider a perturbed problem

$$y' = f(x, y) + \lambda g(x, y), \quad y(x_0) = \eta$$

involving a small perturbation parameter λ, but with the same initial value. Assuming all functions involved to be C^∞ we know that the solution $y(\lambda, x)$ of the perturbed problem is a C^∞ function of λ, and it reduces to $y^{(0)}(x)$ when $\lambda = 0$. It therefore has a Maclaurin series

$$y(\lambda, x) \sim y^{(0)}(x) + \lambda y^{(1)}(x) + \lambda^2 y^{(2)}(x) + \cdots$$

which, like all Maclaurin series, represents the function $y(\lambda, x)$ asymptotically as λ tends to 0. Note that the superscripts in parentheses are supposed to distinguish different vector functions of x; they do not denote derivatives.

It turns out that the coefficient functions $y^{(1)}(x)$, $y^{(2)}(x)$, and so on, can be found recursively by solving a sequence of linear equations. We shall set out the calculation of $y^{(1)}(x)$ in the manner of practical mathematics. The reader should understand that the justification is found in the inductive use of Proposition 6.7, as described in the previous subsection. If we differentiate with respect to λ alone, not varying η, the equation of variation becomes, in a hopefully obvious notation:

$$\frac{d}{dx}\left(\frac{\partial y}{\partial \lambda}\right) = \Big(D_y f(x, y) + \lambda D_y g(x, y)\Big)\left(\frac{\partial y}{\partial \lambda}\right) + g(x, y)$$

with initial condition

$$\left(\frac{\partial y}{\partial \lambda}\right)_{x=x_0} = 0.$$

Setting $\lambda = 0$ we obtain

$$\frac{d}{dx} y^{(1)}(x) = D_y f\big(x, y^{(0)}(x)\big) y^{(1)}(x) + g\big(x, y^{(0)}(x)\big), \quad y^{(1)}(x_0) = 0.$$

Exercise Find the initial value problem satisfied by $y^{(2)}(x)$.

All of this extends to the case where the parameter λ is an r-vector $(\lambda_1, \ldots, \lambda_r)$. An important case of this is a system involving a *complex dependent variable* $y \in \mathbb{C}^n$ and a *complex parameter:*

$$\frac{dy}{dx} = f(x, y, \ell).$$

The same argument that handled the case of a real parameter can be applied here. We can study the dependence of the solution on ℓ by appending the equation

$$\frac{d\ell}{dx} = 0.$$

We deduce that if $f(x, y, \ell)$ is a complex analytic function of the pair $(y, \ell) \in \mathbb{C}^n \times \mathbb{C}$, then the solution of the initial value problem

$$\frac{dy}{dx} = f(x, y, \ell), \quad y(x_0) = \eta, \quad (\ell \in \mathbb{C})$$

is an analytic function of ℓ, as well as of η. This implies, for example, that for fixed η, and knowing a solution for $\ell = \ell_0$ on the interval $]a, b[$, then if $a < a' < x_0 < b' < b$, the solution for ℓ near to ℓ_0 can be expanded in a power series

$$y(x, \ell) = \sum_{k=0}^{\infty} c_k(x)(\ell - \ell_0)^k$$

convergent for $a' \leq x \leq b'$ and $|\ell - \ell_0| < r$, for some $r > 0$.

An important application of this is to linear equations depending on a complex parameter, and will be the key to studying eigenvalue problems in Chap. 7. As an illustration, consider a second order problem

$$y'' + p(x)y' + (q(x) - \ell)y = 0, \quad (a < x < b) \tag{6.7}$$

where $p(x)$ and $q(x)$ are continuous in the interval $]a, b[$ and ℓ is a complex parameter. Clearly we can regard y as complex, so that the direction field of the equivalent first order system

$$\begin{aligned} y_1' &= y_2 \\ y_2' &= -(q(x) - \ell)y_1 - p(x)y_2 \end{aligned}$$

is a complex analytic function of (y_1, y_2, ℓ). Now let $a < c < b$, and let $\phi_1(x, \ell)$ and $\phi_2(x, \ell)$ be solutions of (6.7), such that

$$\phi_1(c, \ell) = 1, \quad \phi_1'(c, \ell) = 0, \quad \phi_2(c, \ell) = 0, \quad \phi_2'(c, \ell) = 1.$$

Then $\phi_1(x, \ell)$ and $\phi_2(x, \ell)$ constitute a solution basis, and are entire complex analytic functions of ℓ. A further deduction on the basis of Proposition 6.6 will be needed in Sect. 7.2: the functions ϕ_1 and ϕ_2, and their partial derivatives with respect to x up to order 2, are continuous functions of the pair (x, ℓ).

6.2.4 Exercises

1. The fact that the solution of the initial value problem is a continuous function of the initial values can have surprising consequences in the realm of practical mathematics. Consider a frictionless simple pendulum of length ℓ (see Sect. 2.3 Exercise 5). This is described by the equation

$$\theta'' = -\frac{g}{\ell} \sin\theta.$$

For initial values $\theta(0) = \alpha$, $\theta'(0) = 0$, where $0 < \alpha < \pi$, the motion consists of oscillations with amplitude α. Let the corresponding period be T_α. Show that $\lim_{\alpha\to\pi-} T_\alpha = \infty$. It follows that for α just short of, but sufficiently near to π, the pendulum oscillates with period that exceeds 1 year.
Hint Apply Proposition 6.3 to the solutions of the system $\theta' = v$, $v' = -(g/L)\sin\theta$ that pass close to the equilibrium point $\theta = \pi$, $v = 0$.
Note The apparent paradox is resolved by noting that a frictionless pendulum is an ideal object that can only be approximated to, but is unachievable in practice. A friction term can be introduced by writing $\theta'' = -\frac{g}{\ell}\sin\theta - \varepsilon\theta'$. The small

positive constant ε has the dimensions of $(\text{time})^{-1}$. The unperturbed solution will be unreliable on a timescale of ε^{-1}.

2. Calculate the solutions ϕ_1, ϕ_2 that satisfy

$$\phi_1(0) = 1,\ \phi_1'(0) = 0,\ \phi_2(0) = 0,\ \phi_2'(0) = 1,$$

for the following problems. Check that they are entire analytic functions of the complex parameter ℓ.

(a) $y'' - \ell y = 0$
(b) $y'' - y' + \ell y = 0$

3. An engineer working at the geothermal energy plant Hellisheidarvirkjun (64° north) drops their spanner down a vertical borehole. The borehole is 250 m deep. How far is the spanner deflected sideways by the Coriolis force?
Hint Let $\mathbf{k}$ be a unit vector parallel to the earth's axis. A coordinate system fixed relative to the earth rotates with angular velocity $\Omega\mathbf{k}$ where $\Omega = 7.3 \times 10^{-5}\,\text{s}^{-1}$. The equation of motion (using vectors in such a coordinate system) of an object near the earth's surface and moving under gravity can be written using the vector product (cross product) of two vectors as:

$$\mathbf{x}'' = \mathbf{g} + 2\,\mathbf{x}' \times \Omega\mathbf{k}$$

where the vector $\mathbf{g}$ is gravitational force, which points to the centre of the earth and has magnitude $10\,\text{m/s}^2$. View Ω as a small parameter and solve the equation of variation to get a first approximation.
Hint Take the origin of coordinates at the location of the plant. Then $\mathbf{g} = -10\mathbf{n}$, where $\mathbf{n}$ is a unit vector directed vertically up, and the unperturbed solution is $\mathbf{x}_0(t) = -5t^2\mathbf{n}$. Compute $\partial\mathbf{x}/\partial\Omega$ at $\Omega = 0$ using the equation of variation.

4. Oscillations of a pendulum with amplitude α satisfy the initial value problem (with reference to exercise 1 we have put $g/\ell = 1$ and replaced θ by x):

$$\frac{d^2x}{dt^2} + \sin x = 0, \quad x(0) = \alpha, \quad \frac{dx}{dt}(0) = 0$$

(a) Let the solution be $x = x(t, \alpha)$. Expand in powers of α:

$$x(t, \alpha) = x_0(t) + \alpha x_1(t) + \alpha^2 x_2(t) + \alpha^3 x_3(t) + O(\alpha^4)$$

and find explicit formulas for $x_0(t)$, $x_1(t)$, $x_2(t)$ and $x_3(t)$.
Hint Time to revisit Sect. 1.4 Exercise 1.

(b) Denoting the period of oscillation by $T(\alpha)$ show that

$$T(\alpha) = 2\pi + c\alpha^2 + O(\alpha^3)$$

where

$$c = -\frac{x_3'(2\pi)}{x_1''(2\pi)}.$$

If you have the patience you can calculate c and compare the result with the more detailed information in Sect. 2.3 Exercise 7(f).
Hint For small α the period is the root of $\frac{\partial x}{\partial t}(t, \alpha) = 0$ near to $t = 2\pi$. This can be expanded in powers of α (justified by the implicit function theorem) and the coefficients of its Maclaurin series found by implicit differentiaton.

5. A planet orbiting the sun with angular momentum per unit mass h, and subject to Newton's laws, follows an orbit that can be described by the differential equation (see Sect. 2.4)

$$\frac{d^2u}{d\theta^2} = -u + \frac{k}{h^2} \tag{6.8}$$

Here we use polar coordinates r and θ, with origin at the sun. The variable u appearing in the equation is $1/r$ and the acceleration due to the sun's gravity is ku^2. General relativity leads to a correction to this equation, namely,

$$\frac{d^2u}{d\theta^2} = -u + \frac{k}{h^2} + 3Mu^2 \tag{6.9}$$

where M is the sun's gravitational mass. Treating M as a small parameter (see the discussion below), show that, to the first order in M, a solution $u = U(\theta)$ of (6.8), satisfying some initial conditions, should be upgraded to

$$u = U(\theta) + MV(\theta)$$

where $v = V(\theta)$ is the solution of the linear inhomogeneous initial value problem

$$\frac{d^2v}{d\theta^2} + v = 3U(\theta)^2, \quad v(0) = 0, \ v'(0) = 0$$

Note 1 If this calculation is followed through for the planet Mercury, one obtains the first order relativistic correction to the Keplerian elliptic orbit. The corrected orbit can be viewed as an ellipse that rotates about its centre at the rate of 43 s of arc per century. This is close enough to the residual unexplained value, known previously from observation, that it convinced Einstein that his theory was correct.
Note 2 Gravitational mass, in relativity, is measured in distance units. The mass of the sun is 1500 m. Compared to the semi-major axis of the orbit of Mercury, 5.8×10^{10} m, this looks like a small parameter. This is reinforced by other considerations. For example, if we are studying the orbit of the earth we can

introduce a new, dimensionless variable $\tilde{u} = Ru$, where R is the mean radius of the earth's orbit. We then obtain the equation

$$\frac{d^2\tilde{u}}{d\theta^2} = -\tilde{u} + \frac{Rk}{h^2} + \frac{3M\tilde{u}^2}{R}.$$

For the earth we have, approximately, $\tilde{u} \approx 1$, $Rk/h^2 \approx 1$, $M/R \approx 10^{-8}$. So it makes sense to expand $\tilde{u}$ in powers of M/R. This is the same as expanding u in powers of M.

6. Let $\phi_{x_0}^{x}(\eta)$ be the solution of the initial value problem $y' = f(x, y)$, $y(x_0) = \eta$. We assume that $f : \mathbb{R} \times \mathbb{R}^n \to \mathbb{R}^n$, is continuous and satisfies condition L, and that all solutions are maximally extended.

   (a) Suppose that for all x_0 in an interval $]a, b[$ and for all η, the solution can be extended to at least the whole of $]a, b[$. Show that

   $$\phi_{x_2}^{x_3} \circ \phi_{x_1}^{x_2} = \phi_{x_1}^{x_3}$$

   for all x_1, x_2 and x_3 in $]a, b[$.

   (b) Let $A \subset \mathbb{R}^n$, let $x_1 > x_0$, and suppose that all solutions of the initial value problem $y' = f(x, y)$, $y(x_0) = \eta$, with $\eta \in A$, can be extended beyond $x = x_1$. Show that $\phi_{x_0}^{x_1}$ is a homeomorphism of A on to $\phi_{x_0}^{x_1}(A)$, its inverse being $\phi_{x_1}^{x_0}$. Show, moreover, that if A is open, so is $\phi_{x_0}^{x_1}(A)$.

   (c) In the context of the last item we suppose further that f is a C^1 function and A is open. Show that $\phi_{x_0}^{x_1}$ is a diffeomorphism of A on to $\phi_{x_0}^{x_1}(A)$. Show further that

   $$\left(D\phi_{x_0}^{x_1}(\eta)\right)^{-1} = (D\phi_{x_1}^{x_0})\left(\phi_{x_0}^{x_1}(\eta)\right)$$

   (on the right is the derivative of $\phi_{x_1}^{x_0}$ computed at the point $\phi_{x_0}^{x}(\eta)$).

   (d) Suppose that f is independent of x (that is, the equation is autonomous; it is of the form $y' = f(y)$). Show that

   $$\phi_{x_0}^{x} = \phi_0^{x-x_0}$$

   (e) Suppose that all solutions of the autonomous equation $y' = f(y)$ (posed in $\mathbb{R}^n$) can be extended to the whole real line $\mathbb{R}$. Let $\Phi^x := \phi_0^x$. Show that

   $$\Phi^{x_1} \circ \Phi^{x_2} = \Phi^{x_1+x_2}.$$

   Hint The mapping $(x, \eta) \mapsto \Phi^x(\eta)$ is a differentiable action of the additive group $(\mathbb{R}, +)$ on the space $\mathbb{R}^n$. Such an action is often called a *flow*.

7. Let D be a domain of the form $\mathbb{R} \times M$, where $M \subset \mathbb{R}^n$ and is open. Assume that $f(x, y)$, where $f : D \to \mathbb{R}^n$, is periodic in x with period T. Assume also that f is continuous and satisfies condition L.

(a) Show that $\phi(x)$ is a solution of $y' = f(x, y)$ if and only if $\phi(x+T)$ is a solution.
(b) Show that

$$\phi_{x_1+T}^{x_2+T} = \phi_{x_1}^{x_2}.$$

(c) Let $\eta \in M$. Show that the solution that satisfies $y(0) = \eta$ is periodic with period T (though T is not necessarily the least period) if and only if $\phi_0^T(\eta) = \eta$.

8. Suppose that f is a C^1 function in the domain D.

(a) Prove that, as long as the solution through (x_0, η) can be extended beyond x, then

$$\frac{d}{dx} \det D\phi_{x_0}^x(\eta) = \big(\mathrm{div}_y f\big)\big(x, \phi_{x_0}^x(\eta)\big) \det D\phi_{x_0}^x(\eta)$$

Hint See Sect. 5.2 Exercise A1.
(b) Deduce that

$$\det D\phi_{x_0}^x(\eta) = \exp\left(\int_{x_0}^x \mathrm{div}_y f(s, \phi_{x_0}^s(\eta))\, ds\right).$$

(c) Show that if $\mathrm{div}_y f = 0$ throughout D then the diffeomorphisms $\phi_{x_0}^x$ preserve n-dimensional volume in $\mathbb{R}^n$.

Note 1 The divergence here is meant to be taken with respect to y alone, that is,

$$\mathrm{div}_y f = \sum_{k=1}^n \frac{\partial f_k}{\partial y_k}.$$

Note 2 Readers who have studied classical mechanics may recognise in item (c) Liouville's theorem, that the phase flow of a Hamiltonian system preserves the volume in phase space.
Note 3 Readers who have studied fluid dynamics may recognise here the criterion for an incompressible fluid flow, that the divergence of the velocity field is zero.

6.3 Essential Stability Theory

In Sect. 2.3, on Newton's equation, we had occasion to distinguish two types of equilibrium points (or constant solutions), designated *stable* and *unstable*. Later, in Sect. 2.4, on motion in a central force field, we observed that a periodic orbit could be stable or unstable. In this section we shall study the notion of stability more closely in connection with constant solutions (equilibrium points) and present precise definitions. We shall obtain important and practical criteria for stability. In keeping with the general philosophy of this text, the results are essential, but not necessarily elementary. In fact, certain parts of the treatment may appear quite demanding.

6.3.1 Stability of Equilibrium Points

Let D be an open connected subset of $\mathbb{R}^n$. We shall consider an equation $dx/dt = F(t, x)$, defined in a set $I \times D$, where I is an interval that includes the interval $[0, \infty[$, and in which the function $F : I \times D \to \mathbb{R}^n$ is C^1. The equilibrium solutions, or constant solutions, are then the points c such that $F(t, c) = 0$ for all $t \in I$. We are using a notation similar to that adopted in Sect. 2.3, where the dependent variable is written as x (though now a point in $\mathbb{R}^n$) and the independent variable is t. It is a temptation, to be encouraged, to think of t as time. The assumption that I includes the interval $[0, \infty[$ reflects our principal interest in future time, and more particularly in the limit $t \to \infty$.

Definition A constant solution c (or equilibrium point) is said to be *stable* if the following condition is satisfied: for each neighbourhood U of c, there exists a neighbourhood V of c, such that for all η in V, the solution $x(t)$ that satisfies $x(0) = \eta$ exists for all $t \geq 0$, and for all such t we have $x(t) \in U$.

Definition A constant solution c is said to be *asymptotically stable* if it is stable and, in addition, the following condition is satisfied: there exists a neighbourhood U of c, such that for all η in U, the solution $x(t)$ that satisfies $x(0) = \eta$ exists for all $t \geq 0$, and $\lim_{t\to\infty} x(t) = c$.

The two conditions defining asymptotic stability, the requirement of stability and the requirement that $\lim_{t\to\infty} x(t) = c$, are independent of each other. In proofs of asymptotic stability one has to pay careful attention to both. One should also note that the terminology varies somewhat between sources. This can be rather confusing. However, the two stability notions defined above are of fundamental importance, even if their names may vary. At the risk of being pedantic, we should state that an equilibrium point that is not stable is said to be unstable.

From now on in this section, until the project following the exercises, we shall consider only the autonomous case, when $F(t, x)$ is independent of t. So F is a C^1 function defined in the set D.

Example Let $F : \mathbb{R} \to \mathbb{R}$ be a C^1 function and let c be an isolated zero of F, in fact we suppose that c is the only zero in the interval $]c - h, c + h[$. The reader should now check the following. If F is positive in $]c - h, c[$ and negative in $]c, c + h[$, then c is an asymptotically stable equilibrium point. If F is negative in $]c - h, c[$ or (no misprint!) positive in $]c, c+h[$, then c is an unstable equilibrium point. Compare the discussion of Sect. 2.1.

The following proposition presents the best known, and probably the most useful, criterion for asymptotic stability of an equilibrium point.

Proposition 6.10 *Let the function $F : D \to \mathbb{R}^n$ be C^1, let $c \in D$ be such that $F(c) = 0$ and let A be the matrix $DF(c)$. If all eigenvalues of A (including complex ones) have negative real parts, then the equilibrium solution $x(t) = c$ is asymptotically stable.*

Proof To simplify the notation we assume that $c = 0$. This can be accomplished by a translation of the coordinates. We write

$$F(x) = Ax + g(x),$$

where A is $DF(0)$ and $g(x)$ is a higher order term, meaning that

$$\lim_{x \to 0} g(x)/|x| = 0.$$

For a given $\eta \in D$ we consider the solution $x(t)$, that satisfies $x(0) = \eta$ to be defined for all t in the maximum interval of existence of the solution; the interval depends in general on η. As long as the solution exists it satisfies

$$\frac{d}{dt}x(t) = Ax(t) + g(x(t))$$

and hence also

$$x(t) = e^{tA}\eta + \int_0^t e^{(t-s)A} g(x(s))\, ds \tag{6.10}$$

by (5.6).

Assume that all eigenvalues of A have negative real parts. By Proposition 5.3, there exist $C \geq 1$ and $m > 0$, depending only on A and not on η, such that

$$|e^{tA}\eta| \leq Ce^{-mt}|\eta|, \quad (t \geq 0,\ \eta \in \mathbb{R}^n).$$

Hence, by (6.10)

$$|x(t)| \leq Ce^{-mt}|\eta| + C\int_0^t e^{-m(t-s)}|g(x(s)|\, ds$$

for all $t > 0$ to which the solution can be extended.

Let $\alpha = m/2C$. There exists $\beta > 0$, such that $|x| < \beta$ implies that $|g(x)| < \alpha|x|$. Let $\delta = \beta/2C$ and consider an initial value η such that $|\eta| < \delta$. Let $\tau > 0$ and suppose that the solution can be extended up to $t = \tau$ and satisfies $|x(t)| < \beta$ for $0 \le t \le \tau$. Then

$$e^{mt}|x(t)| \le C|\eta| + C\alpha \int_0^t e^{ms}|x(s)|\,ds, \quad (0 \le t \le \tau)$$

and by Gronwall's inequality (Lemma 4.3) this implies

$$e^{mt}|x(t)| \le C|\eta|e^{C\alpha t}, \quad (0 < t < \tau).$$

We therefore find that

$$|x(t)| \le C|\eta|e^{(C\alpha - m)t} = C|\eta|e^{-mt/2}, \quad (0 < t < \tau)$$

and, in particular,

$$|\phi^t(\eta)| \le C|\eta| < \beta/2, \quad (0 < t < \tau).$$

The extension principle (Proposition 6.2 and the discussion following it) now allows us to conclude that the solution can be extended for all $t > 0$ and satisfies $|x(t)| < \beta$. By the above calculations it also satisfies

$$|x(t)| \le C|\eta|e^{-mt/2}, \quad (0 < t < \infty).$$

We conclude that if $|\eta| < \delta$ then $\lim_{t\to\infty} x(t) = 0$. This is one requirement of asymptotic stability. To prove stability let $\varepsilon > 0$. If $|\eta| < \min(\delta, \varepsilon/C)$ then

$$|x(t)| \le C|\eta| < \varepsilon$$

for all $t > 0$. This concludes the proof of asymptotic stability. □

If, in the context of Proposition 6.10, the matrix A has an eigenvalue with positive real part, then the equilibrium point is unstable. This will be proved shortly. However, if A has an eigenvalue on the imaginary axis and none with positive real part, then no determination of stability or instability is possible on the basis of A alone. We have to consider higher order terms. More precisely, since $F(c) = 0$ and A is the Jacobian matrix of F at $x = c$, we have

$$F(x) = A(x - c) + G(x)$$

where $\lim_{x\to c} G(x)/|x - c| = 0$. We have to take the function $G(x)$ into account.

We next turn to methods that can often be applied in cases when Proposition 6.10 may be inapplicable.

6.3.2 *Lyapunov Functions*

An important approach to studying the stability of equilibrium points is the use of Lyapunov functions (sometimes called Lyapunov's direct method). This takes its cue from the energy of a mechanical system, which decreases with time along a solution curve in the presence of dissipative forces (such as friction). Again we shall consider an autonomous equation $dx/dt = F(x)$ in a phase space $D \subset \mathbb{R}^n$, in which F is a C^1 mapping.

It is useful to use the notation $\phi^t(\eta)$ to denote the fully extended solution of the initial value problem

$$dx/dt = F(x), \quad x(0) = \eta$$

for a given $\eta \in D$. Then the solution of the initial value problem

$$dx/dt = F(x), \quad x(t_0) = \eta$$

is $\phi^{t-t_0}(\eta)$. We have the important formula, referred to as the *group property*:

$$\phi^{t_1+t_2}(\eta) = \phi^{t_1}(\phi^{t_2}(\eta)).$$

In the case when all solutions can be extended to the whole line $\mathbb{R}$, we can view ϕ^t as a mapping from D to itself.

Exercise In the case just defined, show that the mapping ϕ^t is a C^1 diffeomorphism of D on to itself. Readers that have studied group actions will recognise that the family of mappings $(\phi^t)_{t\in\mathbb{R}}$ defines an action of the additive group $(\mathbb{R}, +)$ on the phase space D. This is often called a *flow* in D. Compare Sect. 6.2 Exercise 6.

As usual when we have an autonomous equation we can concentrate on phase curves rather than solution curves. Thus given $\eta \in D$, the phase curve through η is the *parametrised curve* $x = \phi^t(\eta)$, viewed as a curve in D parametrised by t. The solution curve is the *graph* $\{(t, x) : x = \phi^t(\eta)\}$, viewed as a curve in $\mathbb{R} \times D$.

It is not easy to give a general definition of Lyapunov function. In the propositions presented below, conclusions about the stability, or lack thereof, of an equilibrium point, are obtained by considering the behaviour of a function, defined in the phase space, along phase curves. In all these cases, the function in question may be referred to as a *Lyapunov function for the problem.*

One notion, that of positive definiteness, is commonly encountered in connection with Lyapunov functions, although we do not consider it a necessary property for using the appellation.

Definition Let $W \subset \mathbb{R}^n$ be such that $0 \in W$ and let $V : W \to \mathbb{R}$. The function V is said to be *positive definite* if $V(x) > 0$ for all $x \in W \setminus \{0\}$, but $V(0) = 0$.

Proposition 6.11 *Let the function $F : D \to \mathbb{R}^n$ be C^1 and let $c \in D$ be a zero of F. Let W be a compact neighbourhood of c, and let $V : W \to \mathbb{R}$ be a continuous function with the following properties:*

(i) The function V is non-increasing along phase curves.
(ii) The translated function $x \mapsto V(c + x)$ is positive definite in the translated neighbourhood $W - c$.

Then c is a stable equilibrium point.

Some further explanation might be helpful in advance of the proof. Property (i) means for each $\eta \in D$ the function $t \mapsto V(\phi^t(\eta))$ non-increasing (that is, it is decreasing, though not necessarily strictly) for all $t > 0$ within the interval of existence. This includes the important case when it is constant. Property (ii) simply says that the restriction of V to W is positive except at c, where it attains its minimum, which is 0. However, the phrasing in terms of positive definiteness is conventional (very often it is assumed that c is the origin 0) and the reader will almost certainly encounter it in other texts.

Proof By analysis, positive definiteness implies that the family of sublevel sets

$$H_r := \{x \in W : V(x) < r\}, \quad (r > 0)$$

forms a neighbourhood basis of the point c in $\mathbb{R}^n$. In order to establish stability we may restrict our attention to these neighbourhoods alone. We choose $r > 0$ so that r is strictly less than the minimum of V on the boundary of W (the reader may check this point; it is why we want W to be compact). A phase curve with initial value in H_r cannot therefore leave W; it can therefore be extended for all $t > 0$ (Proposition 6.2). Moreover, since V is non-increasing along phase curves, it remains in H_r for all $t > 0$. This proves stability. □

A slight strengthening of condition (i) (read with care!) yields the following, much used, criterion for asymptotic stability:

Proposition 6.12 *Let the function $F : D \to \mathbb{R}^n$ be C^1 and let $c \in D$ be an isolated zero of F. Let W be a compact neighbourhood of c, containing no zero of F other than c, and let $V : W \to \mathbb{R}$ be a continuous function with the following properties:*

i) The function V is strictly decreasing along non-constant phase curves.
(ii) The translated function $x \mapsto V(c + x)$ is positive definite in the translated neighbourhood $W - c$.

Then c is an asymptotically stable equilibrium point.

The proof uses a concept from dynamical systems theory (a topic for further studies).

Definition Suppose that the solution with initial value η can be extended to all $t > 0$. The *ω-limit set of η*, denoted by $\omega(\eta)$, consists of all $x \in D$ with the following property: there exists an increasing sequence $(t_k)_{k=1}^{\infty}$, such that $t_k \to \infty$, and $\phi^{t_k}(\eta) \to x$.

The basic properties of ω-limit sets needed for the proof of Proposition 6.12 are set out in the following lemmas, the proofs of which are mostly left to the reader.

Lemma 6.1 *If the forward phase curve of η (the part with $t > 0$) lies in a compact subset of D, then $\omega(\eta)$ is not empty.*

Hint for Proof A compact set is sequentially compact.

Lemma 6.2 *If $\xi \in \omega(\eta)$ then both the forward and backward phase curves through ξ lie in $\omega(\eta)$.*

Hint for Proof You will need the group property of the family ϕ^t, $(t \in \mathbb{R})$, explained earlier in this section, and the continuity of ϕ^t (Proposition 6.4).

Lemma 6.3 *Suppose that the forward phase curve of η lies in a compact subset of D. If $\omega(\eta)$ consists of a single point p, then $\lim_{t\to\infty} \phi^t(\eta) = p$.*

Hint for Proof Prove the contrapositive. Actually the assumption that the phase curve lies in a compact set is unnecessary; see the exercises.

Proof of Proposition 6.12 The equilibrium point c is stable by Proposition 6.11. Choose $r > 0$ so that H_r (the sublevel sets were defined in the proof of Proposition 6.11) does not meet the boundary of W. Let $\xi \in H_r$. Then the phase curve $\phi^t(\xi)$ can be extended for all $t > 0$ and does not exit H_r. Therefore, $\omega(\xi)$ is not empty (Lemma 6.1). Moreover, by condition (i), the limit $\lambda := \lim_{t\to\infty} V(\phi^t(\xi))$ exists and $\lambda \geq 0$. Let $p \in \omega(\xi)$ and let t_k be an increasing sequence such that $\phi^{t_k}(\xi) \to p$. We deduce that $V(p) = \lambda$. It follows that V is constant on the set $\omega(\xi)$, with value λ. But a phase curve with initial value in $\omega(\xi)$ remains in $\omega(\xi)$ (Lemma 6.2), and so V is constant along such a phase curve. Therefore $\omega(\xi)$ consists entirely of equilibrium points (by condition (i)). Again, because, $\omega(\xi) \subset W$ and W contains only one equilibrium point, we must have $\omega(\xi) = \{c\}$. By Lemma 6.3, $\phi^t(\xi) \to c$. □

A variation of these ideas produces a criterion for instability.

Proposition 6.13 *Let the function $F : D \to \mathbb{R}^n$ be C^1 and let $c \in D$ be an isolated zero of F. Let W be a neighbourhood of c, containing no zero of F other than c, and let $V : W \to \mathbb{R}$ be a continuous function with the following properties:*

(i) $V(c) = 0$ and the function V is strictly increasing along non-constant phase curves.
(ii) Every neighbourhood of c contains a point at which V is strictly positive.

Then c is an unstable equilibrium point.

Proof Let B be a closed ball with centre c, chosen so that it contains no equilibrium point except c, and $B \subset W$. Let $\xi \in B$ be such that $V(\xi) > 0$. We claim that the forward phase curve $\phi^t(\xi)$, $(t > 0)$, must exit B. Suppose, if possible that it remains in B. Then the phase curve can be extended to all $t > 0$, so that $\omega(\xi)$ is not empty and $\omega(\xi) \subset B$. As in the proof of Proposition 6.12, the function V is a constant λ on $\omega(\xi)$, but in the present case $\lambda > 0$. By condition (i), $\omega(\xi)$ consists entirely of equilibrium points. Hence $\omega(\xi) = \{c\}$. But that contradicts the assumption that $V(c) = 0$. □

Practicalities

In practice, V is almost always a C^1 function. The growth or decay of V along a phase curve can then be studied by calculating the derivative:

$$\frac{d}{dt} V(\phi^t(\eta)) = \nabla V(\phi^t(\eta)) \cdot F(\phi^t(\eta)).$$

We immediately deduce the following criteria, which greatly aid in the application of the propositions of this section:

1. If $\nabla V(x) \cdot F(x) < 0$ for all x, except when $F(x) = 0$, then V is strictly decreasing along non-constant phase curves.
2. If $\nabla V(x) \cdot F(x) \leq 0$ for all x, then V is non-increasing along phase curves.
3. If $\nabla V(x) \cdot F(x) > 0$ for all x, except when $F(x) = 0$, then V is strictly increasing along non-constant phase curves.

We conclude with a couple of worked examples, which pave the way for the more general discussion of the next section.

Example 1 We consider the plane system, written using coordinates x and y as:

$$\frac{dx}{dt} = -x + f(x, y), \qquad \frac{dy}{dt} = -y + g(x, y)$$

where f and g are C^1 functions such that f, g, $\partial f/\partial x$, $\partial f/\partial y$, $\partial g/\partial x$ and $\partial g/\partial y$ are all 0 for $x = y = 0$.

We let $V(x, y) = x^2 + y^2$. The positive definiteness of V is evident. Moreover, going over to polar coordinates from the third line, we have:

$$\begin{aligned}
\nabla V(x, y) &\cdot (-x + f(x, y), -y + g(x, y)) \\
&= -2x^2 + 2xf(x, y) - 2y^2 + 2yg(x, y) \\
&= -2r^2 + 2r\cos\theta f(r\cos\theta, r\sin\theta) + 2r\sin\theta g(r\cos\theta, r\sin\theta) \\
&= -2r^2\Big(1 - r^{-1}\cos\theta f(r\cos\theta, r\sin\theta) - r^{-1}\sin\theta g(r\cos\theta, r\sin\theta)\Big)
\end{aligned}$$

Since f, g and their partial derivatives are all 0 at $x = y = 0$, there exists $\alpha > 0$, such that the quantity enclosed in large parentheses is greater than $\frac{1}{2}$ when $0 < r < \alpha$. So for r in this range the function V decreases strictly along phase curves. By the inverse function theorem the origin is an isolated equilibrium point, and we can choose α so that there is no equilibrium point for which $0 < r < \alpha$. It follows by Proposition 6.12 that the equilibrium point $(0, 0)$ is asymptotically stable, a conclusion that can also be reached by applying Proposition 6.10.

Example 2 We consider the plane system:

$$\frac{dx}{dt} = -x + f(x, y), \quad \frac{dy}{dt} = y + g(x, y)$$

where f and g are C^1 functions such that f, g, $\partial f/\partial x$, $\partial f/\partial y$, $\partial g/\partial x$ and $\partial g/\partial y$ are all 0 for $x = y = 0$.

We let $V(x, y) = -x^2 + y^2$. Then:

$$\begin{aligned}
&\nabla V(x, y) \cdot (-x + f(x, y), y + g(x, y)) \\
&\qquad = 2x^2 - 2xf(x, y) + 2y^2 + 2yg(x, y) \\
&\qquad = 2r^2 - 2r\cos\theta f(r\cos\theta, r\sin\theta) + 2r\sin\theta g(r\cos\theta, r\sin\theta) \\
&\qquad = 2r^2\Big(1 - r^{-1}\cos\theta f(r\cos\theta, r\sin\theta) + r^{-1}\sin\theta g(r\cos\theta, r\sin\theta)\Big)
\end{aligned}$$

As in Example 1 there exists $\alpha > 0$, such that the quantity enclosed in large parentheses is greater than $\frac{1}{2}$ when $0 < r < \alpha$, and no equilibrium point exists within this range. For r in this range the function V increases strictly along phase curves, and $V(0, \beta) > 0$ for all $\beta > 0$. It follows by Proposition 6.13 that the equilibrium point $(0, 0)$ is unstable.

6.3.3 *Construction of a Lyapunov Function for the Equation $dx/dt = Ax$*

The culmination of this section is the proof that, in the context of Proposition 6.10, if the matrix A has an eigenvalue with positive real part, then the equilibrium point $x = c$ is unstable. We first reprove Proposition 6.10 using a Lyapunov function. This repetition is justified, firstly, by the fact that the new proof is arguably more geometrical and intuitive, and, secondly, because it prepares the reader for the proof of instability (Proposition 6.14), which is based on similar ideas but is more complicated.

Case 1: All Eigenvalues of A Have Negative Real Part

We shall construct a positive definite quadratic form $Q(x)$, such that

$$\nabla Q(x) \cdot Ax \leq -\sigma Q(x), \quad (x \in \mathbb{R}^n)$$

for some positive constant σ. This shows, by Proposition 6.12, that the equilibrium point $x = 0$ is asymptotically stable. This also follows from Proposition 5.3, but the construction of a Lyapunov function is an approach that extends to a non-linear equation of the form $dx/dt = F(x)$ with $F(0) = 0$ and $DF(0) = A$, as we shall see.

The form $Q(x)$ is given by

$$Q(x) = \sum_{k=1}^{n} z_k \overline{z}_k$$

where $z_1, \ldots, z_n$ are the coordinates, complex in general, of the real vector $x = (x_1, \ldots, x_n) \in \mathbb{R}^n$ in a suitably chosen complex basis $w^{(1)}, \ldots, w^{(n)}$ of the complexification $\mathbb{R}^n + i\mathbb{R}^n$.

Let T be the transition matrix (in general with complex entries) effecting the change of coordinates, so that a vector $x = (x_1, \ldots, x_n) \in \mathbb{R}^n + i\mathbb{R}^n$ is represented in the basis $w^{(1)}, \ldots, w^{(n)}$ by the coordinate vector $(z_1, \ldots, z_n) = \zeta = T^{-1}x$. Then, for a *real* vector x we have:

$$Q(x) = |\zeta|^2 = T^{-1}x \cdot \overline{T}^{-1}x$$

and so

$$\begin{aligned} \nabla Q(x) \cdot Ax &= T^{-1}Ax \cdot \overline{T}^{-1}x + T^{-1}x \cdot \overline{T}^{-1}Ax \\ &= (T^{-1}AT)\zeta \cdot \overline{\zeta} + \zeta \cdot \overline{(T^{-1}AT)}\,\overline{\zeta} \\ &= 2\mathrm{Re}\left((T^{-1}AT)\zeta \cdot \overline{\zeta}\right). \end{aligned}$$

Now, if A is diagonalisable, with eigenvalues $\lambda_1, \ldots, \lambda_n$ (including complex eigenvalues), repeated according to their multiplicities, then we may choose the transition matrix T so that $T^{-1}AT$ is the diagonal matrix

$$\mathrm{diag}(\lambda_1, \ldots, \lambda_n) := \begin{bmatrix} \lambda_1 & 0 & \cdots & 0 \\ 0 & \lambda_2 & \cdots & 0 \\ \vdots & \vdots & \ddots & \vdots \\ 0 & 0 & \cdots & \lambda_n \end{bmatrix}$$

Then we have

$$\nabla Q(x) \cdot Ax = 2 \sum_{k=1}^{n} (\operatorname{Re} \lambda_k) |z_k|^2.$$

If therefore we choose $\sigma > 0$ so that, for all eigenvalues, $\operatorname{Re} \lambda_k < -\frac{1}{2}\sigma < 0$, then

$$\nabla Q(x) \cdot Ax \le -\sigma \sum_{k=1}^{n} |z_k|^2 = -\sigma Q(x)$$

as required.

However, it may be impossible to diagonalise A. No matter, we can *nearly* diagonalise it. As we shall presently see, for a given $\varepsilon > 0$, we can choose the transition matrix T so that

$$T^{-1}AT = \begin{bmatrix} \lambda_1 & \mu_1 & 0 & \cdots & \cdots & 0 \\ 0 & \lambda_2 & \mu_2 & \cdots & \cdots & 0 \\ \vdots & \vdots & \vdots & \ddots & \vdots & \vdots \\ 0 & 0 & 0 & \cdots & \lambda_{n-1} & \mu_{n-1} \\ 0 & 0 & 0 & \cdots & 0 & \lambda_n \end{bmatrix}$$

where, again the eigenvalues are on the main diagonal, each of the quantities μ_k on the superdiagonal (the diagonal above and adjacent to the main diagonal) is either ε or 0, and all other entries are 0. Now it is easy to verify that

$$\nabla Q(x) \cdot Ax \le 2 \sum_{k=1}^{n} (\operatorname{Re} \lambda_k) |z_k|^2 + 2(n-1)\varepsilon |\zeta|^2.$$

The additional term on the right arises from the off-diagonal entries of the matrix $T^{-1}AT$. Now if initially ε is chosen sufficiently small, then we can find $\sigma > 0$ so that

$$2 \max_{1 \le k \le n} \operatorname{Re} \lambda_k + 2(n-1)\varepsilon < -\sigma < 0$$

and from this we conclude

$$\nabla Q(x) \cdot Ax \le -\sigma Q(x).$$

To construct the "near diagonalisation" we first transform A to Jordan canonical form, using a transition matrix T_1:

$$T_1^{-1}AT_1 = \begin{bmatrix} [J_1] & \cdots & \cdots & \cdots \\ \cdots & [J_2] & \cdots & \cdots \\ & & \ddots & \\ \cdots & \cdots & \cdots & [J_p] \end{bmatrix}$$

where each matrix J_k is a Jordan block, one for each distinct eigenvalue, and all other entries are 0. A Jordan block for an eigenvalue λ has the form:

$$\begin{bmatrix} \lambda & 1 & 0 & \cdots & 0 \\ 0 & \lambda & 1 & \cdots & 0 \\ \vdots & \vdots & \ddots & \ddots & \vdots \\ 0 & 0 & 0 & \cdots & 1 \\ 0 & 0 & 0 & \cdots & \lambda \end{bmatrix}$$

Then, using a further, diagonal transition matrix

$$T_2 = \mathrm{diag}(\varepsilon, \varepsilon^2, \ldots, \varepsilon^n)$$

we compute $T_2^{-1}(T_1^{-1}AT_1)T_2$. The reader can check that this converts all appearances of 1 on the superdiagonal of the Jordan canonical form to ε.

The nice thing is that the function $Q(x)$ is also a Lyapunov function for the nonlinear equation

$$x' = Ax + g(x)$$

where g is a higher order term, more precisely, a C^1 function such that $g(x)/|x| \to 0$ as $x \to 0$. For we have

$$\nabla Q(x) \cdot (Ax + g(x)) \leq -\sigma Q(x) + \nabla Q(x) \cdot g(x)$$

and, given $\alpha > 0$

$$|\nabla Q(x) \cdot g(x)| \leq |\nabla Q(x)|\,|g(x)| \leq \alpha |x|^2$$

provided $|x|$ is sufficiently small. This gives

$$\nabla Q(x) \cdot (Ax + g(x)) \leq -\sigma Q(x) + \alpha |x|^2$$

so it is enough to choose α so that

$$Q(x) > \frac{\alpha}{2\sigma}|x|^2$$

for all $x \neq 0$, possible because $Q(x)$ is a positive definite quadratic form. Then α determines a ball with centre 0 in which $Q(x)$ is a positive definite function, strictly decreasing along non-constant phase curves. This gives a second proof of Proposition 6.10.

The use of the quadratic form $Q(x)$ as a Lyapunov function has a pleasing geometric feel. The level sets $Q(x) = C$ (with $C > 0$) form a family of concentric hyperellipsoids; ellipses in two dimensions, ellipsoids in three. Phase curves cross them transversally (without tangency) from outside to inside. For the linear equation $dx/dt = Ax$ this applies to all non-constant phase curves; for the non-linear equation $dx/dt = Ax + g(x)$ to those that begin inside a suitable neighbourhood of 0. This is illustrated in Fig. 6.4.

Case 2: At Least One Eigenvalue of A Has Positive Real Part

Suppose that all eigenvalues $\lambda_1, \ldots, \lambda_m$ have strictly positive real parts, whilst the eigenvalues $\lambda_{m+1}, \ldots, \lambda_n$ have negative or zero real parts. In the case at hand we have $m \geq 1$. Again using the near diagonalisation defined previously, with ε yet to be specified, we set

$$Q_1(x) = \sum_{k=1}^{m} |z_k|^2, \qquad Q_2(x) = \sum_{k=m+1}^{n} |z_k|^2$$

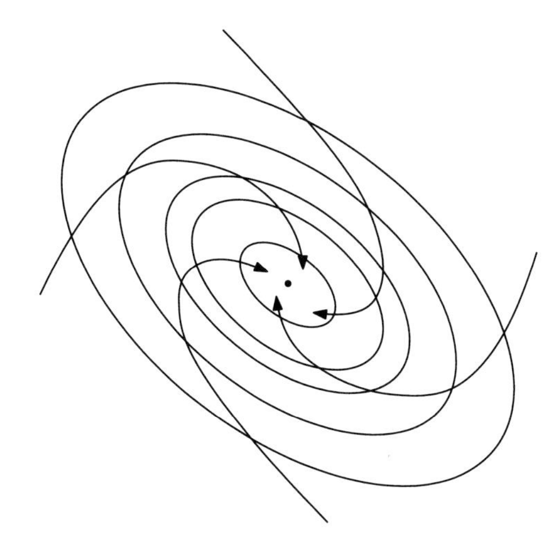

Fig. 6.4 Lyapunov's direct method

where we must understand $\zeta = (z_1, \ldots, z_n) = T^{-1}x$ as the complex coordinate vector of x in the transformed basis. This interpretation is continued in what follows, where 'ζ' is used where the reader might expect 'x'.

We shall deploy as Lyapunov function:

$$Q(x) := Q_1(x) - Q_2(x).$$

We begin by defining some constants that will run through the whole discussion. We let $h = \min_{1 \leq k \leq m} \operatorname{Re} \lambda_k$ and note that $h > 0$. We introduce $C > 0$, such that

$$|\nabla Q_1(x)| \leq C|\zeta|, \quad |\nabla Q_2(x)| \leq C|\zeta|$$

for all x (possible because Q_1 and Q_2 are quadratic forms).

The equilibrium solution $x(t) = 0$ is obviously unstable in the linear case (the reader should reflect on this). Therefore we proceed straight to the non-linear equation

$$\frac{dx}{dt} = Ax + g(x)$$

where $g(x)$ is a C^1 mapping and $\lim_{x \to 0} g(x)/|x| = 0$.

Given $\gamma > 0$, which will be specified later, there is a compact neighbourhood U of 0, such that for all $x \in U$ we have

$$|g(x)| < \gamma|\zeta|$$

Now, for $x \in U$, we have:

$$\begin{aligned} \nabla Q_1(x) \cdot (Ax + g(x)) &\geq 2\sum_{k=1}^{m} (\operatorname{Re} \lambda_k)|z_k|^2 - K\varepsilon \sum_{k=1}^{m} |z_k|^2 - C\gamma|\zeta|^2 \\ &\geq (2h - K\varepsilon - C\gamma)Q_1(x) - C\gamma Q_2(x), \end{aligned}$$

where the positive number K arises from estimating the contribution of the off-diagonal terms of the Jordan canonical form of A, and depends only on the matrix A.

Again, for $x \in U$, we have:

$$\begin{aligned} \nabla Q_2(x) \cdot (Ax + g(x)) &\leq 2 \sum_{k=m+1}^{n} (\operatorname{Re} \lambda_k)|z_k|^2 + K\varepsilon \sum_{k=m+1}^{n} |z_k|^2 + C\gamma|\zeta|^2 \\ &\leq K\varepsilon Q_2(x) + C\gamma|\zeta|^2 = C\gamma Q_1(x) + (K\varepsilon + C\gamma)Q_2(x). \end{aligned}$$

Hence, taking these together, we have

$$\nabla Q(x) \cdot (Ax + g(x)) \geq (2h - K\varepsilon - 2C\gamma)Q_1(x) - (2C\gamma + K\varepsilon)Q_2(x) \qquad (6.11)$$

for all $x \in U$.

We still have to decide on γ and ε. Recall that h was determined by the eigenvalues of A with positive real part, and is positive, and K depends only on the Jordan canonical form of A. However, C depends on Q_1 and Q_2, and they depend on the near diagonalisation, and therefore on ε. And U depends on γ. The order of these dependencies is crucial for the correctness of the following argument.

We first choose $\varepsilon > 0$ so that

$$K\varepsilon < h.$$

This determines the near diagonalisation, hence Q_1 and Q_2, hence C. Then we choose $\gamma > 0$ so that

$$2C\gamma + K\varepsilon < h.$$

Recall that γ determines the neighbourhood U. Next we choose σ so that

$$2C\gamma + K\varepsilon < \sigma < h.$$

These choices could have been specified at the start, but lacking thereby the hindsight arising from inequality (6.11).

The reader may now check using (6.11) that for $x \in U$ we have

$$\nabla Q(x) \cdot (Ax + g(x)) \geq \sigma Q(x),$$

or, in other words, along a phase curve in the neighbourhood U we have

$$\frac{dQ}{dt} \geq \sigma Q.$$

If the phase curve commences at a point η in U, and $Q(\eta) > 0$, then we deduce from this that

$$Q(x(t)) \geq Q(\eta)e^{\sigma t}$$

for as long as the phase curve can be extended and remains in U.

Exercise Prove this claim.

If the phase curve does not exit U then it can be extended for all $t > 0$ by Proposition 6.2. This is incompatible with the exponential lower bound for the growth of $Q(x(t))$. Hence the phase curve must exit U.

The conclusion obtained in the previous paragraph depends on knowing that $Q(\eta) > 0$. There is only one more thing to do, and it is left to the reader:

Exercise Prove that there exist points $\eta \neq 0$, arbitrarily near to 0, at which $Q(\eta) > 0$. It looks obvious from the definition of Q, but there is a catch: we want η to be a *real* vector.

Hint With respect to the transformed complex basis $w^{(1)}, \ldots, w^{(n)}$ an eigenvector of A, whether real or complex, has one coordinate equal to 1 and the rest 0.

We can finally state a proposition, proved by the preceding discussion. It is a companion to Proposition 6.10.

Proposition 6.14 *Let the function $F : D \to \mathbb{R}^n$ be C^1, let $c \in D$ be such that $F(c) = 0$ and let A be the matrix $DF(c)$. If at least one eigenvalue of A has positive real part, then the equilibrium solution $x(t) = c$ is unstable.*

6.3.4 Exercises

1. The plane system

$$\frac{dx}{dt} = x(\lambda - ax - by)$$
$$\frac{dy}{dt} = y(\mu + cx - dy)$$

has been proposed as a model for two competing species: prey, of magnitude x, and predator, of magnitude y. The constants λ, μ, a, b, c and d are all positive.

   (a) Find all equilibrium points on the x and y-axes and show that if $\lambda d > \mu b$ they are both unstable.
   (b) Suppose that $\lambda d > \mu b$. Show that there is a unique equilibrium point (x_0, y_0) with $x_0 > 0$ and $y_0 > 0$, and show that it is asymptotically stable.

2. Show that the function $x^4 + 2y^2$ can be used as a Lyapunov function for the plane systems

$$\text{(i)} \quad x' = y - x^3, \quad y' = -x^3$$
$$\text{(ii)} \quad x' = y + x^3, \quad y' = -x^3$$

and make some deductions about the stability of the equilibrium point $(0, 0)$ in each case.
Note The matrix A (defined by Proposition 6.10) is the same for both systems, but the stability conclusions are different.

3. Find all equilibrium points of the plane system

$$x' = x - y, \quad y' = 4x^2 + 2y^2 - 6$$

and for each, determine whether it is stable, asymptotically stable or unstable.
4. Find a suitable Lyapunov function $V(x, y)$ that can be used to study the equilibrium point $(0, 0)$ of the plane system

$$x' = -x + 9y + g_1(x, y), \quad y' = -y + g_2(x, y).$$

The functions g_1 and g_2 are higher order terms, that is, they and their first order partial derivatives are zero at $x = y = 0$.
5. Consider an autonomous equation $dx/dt = F(x)$, with $F : D \to \mathbb{R}^n$ a C^1 function in the open connected domain $D \subset \mathbb{R}^n$. Suppose the solution with initial value η can be extended for all $t > 0$ and suppose that the limit set $\omega(\eta)$ consists of a single point p. Prove that $\lim_{t\to\infty} \phi^t(\eta) = p$.
Hint Prove the contrapositive. Suppose that the limit does not exist, or it does but is not p. Then there is closed ball with centre p, included entirely in D, which the phase curve repeatedly enters and leaves, thereby crossing the boundary of the ball infinitely often.
Note A similar argument, using more sophisticated metric space theory, can be used to show that if $\omega(\eta)$ has a compact connected component M, then $\omega(\eta) = M$.
6. Let $x = c$ be an equilibrium point of an autonomous equation $x' = f(x)$ in $\mathbb{R}^n$. Let M be the set of all points η such that the phase curve through η converges to c as $t \to \infty$. The set M is called the *stable manifold* of the equilibrium point c.

   (a) Suppose that c is asymptotically stable. Show that M is a connected open set that contains c.
   Hint Some knowledge of metric spaces is needed. Firstly, by asymptotic stability, M includes a ball B with centre c. Then M consists of all η, such that $\phi^t(\eta) \in B$ for some $t > 0$.
   (b) Suppose that there are distinct asymptotically stable critical points c and d. Prove that there must exist a point η such that the phase curve through η converges to neither of them.

7. Consider the plane system

$$x' = y, \quad y' = -x^3 + x - ky$$

where k is a positive constant.

   (a) Show that of the three critical points $(0, 0)$, $(1, 0)$ and $(-1, 0)$, the first is unstable, whilst the second and third are asymptotically stable.
   (b) Show that the function $E(x, y) = \frac{1}{4}x^4 - \frac{1}{2}x^2 + \frac{1}{2}y^2$ is strictly decreasing along non-constant phase curves.

(c) Draw a "map" showing the level curves of $E(x, y)$.
(d) Show that every point in the set

$$N_+ = \{(x, y) : x > 0, \ E(x, y) < 0\}$$

is drawn to $(1, 0)$ as $t \to \infty$ while every point in the set

$$N_- = \{(x, y) : x < 0, \ E(x, y) < 0\}$$

is drawn to $(-1, 0)$.
(e) Try to figure out what the stable manifolds (see Exercise 6) of $(1, 0)$ and $(-1, 0)$ look like and what happens to points on neither of them as $t \to \infty$.

8. Euler's equations for the free rotation of a rigid body, with a point fixed in space, are

$$A \frac{d\Omega_1}{dt} = (B - C)\,\Omega_2\Omega_3$$
$$B \frac{d\Omega_2}{dt} = (C - A)\,\Omega_3\Omega_1$$
$$C \frac{d\Omega_3}{dt} = (A - B)\,\Omega_1\Omega_2.$$

Physicists call $(\Omega_1, \Omega_2, \Omega_3)$ the *angular velocity vector referred to a frame that is fixed in the body as it rotates* and for which *the coordinate axes are the principal axes of inertia*. The quantities A, B, C are positive constants, the *principal moments of inertia*. None of this matters for this problem. However, we assume that $A < B < C$. By a *constant of the motion* we mean a function of $(\Omega_1, \Omega_2, \Omega_3)$ that is constant on phase curves.

(a) Show that the only equilibrium points are the points on the three rays through $(1, 0, 0)$, $(0, 1, 0)$ and $(0, 0, 1)$. These three solutions correspond to the body rotating with constant angular velocity about a principal axis of inertia.
(b) Show that $A\Omega_1^2 + B\Omega_2^2 + C\Omega_3^2$ is a constant of the motion.
(c) Find another constant of the motion of the form $\lambda\Omega_2^2 + \mu\Omega_3^2$. Use it together with item (b) to show that the equilibrium points on the ray through $(1, 0, 0)$ are stable.
(d) Study the stability of $(0, 1, 0)$ and $(0, 0, 1)$ by similar methods.

Note. You can demonstrate the conclusions of (c) and (d) by spinning a mobile phone in the air. Make sure it's an old one.

9. The use of a near diagonalisation is key to the following very useful result:

   Let B be an $n \times n$-matrix of which all eigenvalues λ satisfy $|\lambda| < 1$. Then for all $x \in \mathbb{R}^n$ we have $\lim_{k\to\infty} B^k x = 0$, and the limit is uniform when x is restricted to a bounded set.

   Prove this by constructing a positive definite quadratic form $Q(x)$, such that there exists σ in the range $0 < \sigma < 1$, such that

$$Q(Bx) \leq \sigma Q(x), \quad (x \in \mathbb{R}^n).$$

   Hint Using the same definition of $Q(x)$ as in case 1 in the text (with B instead of A), estimate $Q(Bx)$. It could help to try it first for a diagonalisable matrix. For the convergence to 0 one can observe that $\sqrt{Q(x)}$ is a new norm on $\mathbb{R}^n$.

6.3.5 Projects

A. *Project on stability for periodic differential equations*

The result of Exercise 9 is useful for studying the stability of solutions of time-periodic equations. We shall consider only the simplest case here, that of the zero solution of a homogeneous linear problem. In the following exercises we consider the problem

$$\frac{dx}{dt} = A(t)x \tag{6.12}$$

where the real matrix function $A(t)$ is continuous and periodic in t with period T. We recall the time advance mapping $\Phi_{t_0}^t$ and the period advance mapping $B := \Phi_0^T$. It might help to revisit Sect. 5.3 project A, in particular, Sect. 5.3 Exercises A1 and A2.

A1. Assume that all eigenvalues λ of B satisfy $|\lambda| < 1$.

   (a) Show that every solution $x(t)$ of (6.12) satisfies $\lim_{t\to\infty} x(t) = 0$.
   (b) Show that there exists a constant K such that $|\Phi_0^t(\eta)| \leq K|\eta|$ for all $t > 0$ and all η.
   (c) Show that the solution $x(t) = 0$ is asymptotically stable.

A2. Consider the two-dimensional case of (6.12), that is, $n = 2$ and $A(t)$ is a 2×2-matrix. Suppose further that $\operatorname{tr} A(t) = 0$ for all t.

   (a) Show that $\det B = 1$.
   (b) Suppose that $|\operatorname{tr} B| < 2$. Show that there exists a constant K such that $|\Phi_0^t(\eta)| \leq K|\eta|$ for all $t > 0$ and all η, and deduce that the solution $x(t) = 0$ is stable.

Hint Show that the eigenvalues of B are non-real and on the unit circle. Therefore, there exists an invertible matrix T, such that $T^{-1}BT$ is a rotation of the plane about the origin.
Note Since B preserves area, the solution $x(t) = 0$ cannot be asymptotically stable.

(c) Suppose that $|\text{tr}\, B| > 2$. Show that there exists $\eta \in \mathbb{R}^2$, such that the solution with initial value η is unbounded as $t \to \infty$. Deduce that the solution $x(t) = 0$ is unstable.
Hint Show that the eigenvalues are real, with one inside and the other outside the unit circle.

The previous exercise has an interesting application to the equation of simple harmonic motion with variable frequency, for example a pendulum or swing being "pumped up" by varying its length periodically. We ask: with what period must we vary the length in order that the equilibrium state (with the pendulum hanging downwards) is unstable? We will only consider small oscillations; therefore we pose the linear problem $x'' + \omega(t)x = 0$, where $\omega(t)$ is continuous, positive and periodic with period T. In a phase plane we have the system

$$x' = y, \quad y' = -\omega(t)x.$$

Note that the trace of the matrix is 0, so that the result of Exercise A2 is applicable. It is a challenge to calculate the period advance mapping. Therefore, we simplify by assuming that

$$\omega(t) = \omega_0 + \varepsilon\omega_1(t),$$

where $\omega_0 > 0$ is a constant, ε is a small parameter and $\omega_1(t)$ has period T. Now the period advance mapping depends continuously on ε; we write it as $B(\varepsilon)$.

A3. (a) Consider the unperturbed problem $\varepsilon = 0$. Show that

$$B(0) = \begin{bmatrix} \cos\sqrt{\omega_0}T & \dfrac{1}{\sqrt{\omega_0}}\sin\sqrt{\omega_0}T \\ -\sqrt{\omega_0}\sin\sqrt{\omega_0}T & \cos\sqrt{\omega_0}T \end{bmatrix}$$

Hint Do not worry that the unperturbed equation does not depend on t. We still define B to be the T-advance mapping.

(b) Show that $B(0)$ satisfies $|\text{tr}\, B(0)| < 2$ if and only if T is not an integral multiple of $\pi/\sqrt{\omega_0}$.

(c) Show that if T is not an integral multiple of $\pi/\sqrt{\omega_0}$, then, for sufficiently small ε we have $|\text{tr}\, B(\varepsilon)| < 2$.
Note For such T and ε the solution $x(t) = 0$ is stable and the swing cannot be "pumped up". Of course, how small ε should be depends on T. Note that $\pi/\sqrt{\omega_0}$ is *half the natural period of the swing*. The greatest effect is achieved by "pumping" at this period.

Chapter 7
Sturm-Liouville Theory

... and yes I said yes I will Yes.

In this chapter we consider an important set of classical problems that are of immediate relevance to the theory of Fourier series and the solving of partial differential equations that arise in physics and engineering. We study first and foremost the second order linear problem posed in Sturm-Liouville form.

The homogeneous equation has the form

$$\frac{d}{dx}\Big(p(x)\frac{dy}{dx}\Big) + q(x)y = 0, \quad (a < x < b) \tag{7.1}$$

where the coefficient functions $p(x)$ and $q(x)$ are continuous and real valued in the closed and bounded interval $[a, b]$ and $p(x) \neq 0$ for $a \leq x \leq b$.

If $p(x)$ has a continuous derivative the problem can be written in the standard form for a second order linear equation. But even if it has not we can study existence of solutions by converting it to a linear system in a phase plane. By writing $u = y$ and $v = p(x)y'$ we obtain the equivalent system

$$u' = p(x)^{-1}v, \quad v' = -q(x)u. \tag{7.2}$$

Since, for a given x_0 in $[a, b]$, this has a unique solution in $]a, b[$ for any given initial condition $u(x_0) = c_1$, $v(x_0) = c_2$, we easily find that the solution space of (7.1) is a two-dimensional space of functions defined on $]a, b[$, which extend, together with their first derivatives, to the closed interval $[a, b]$. Moreover, two solutions $y_1(x)$ and $y_2(x)$ of (7.1) form a basis for the solution space if and only if the corresponding vectors of initial conditions at $x = x_0$ in the phase plane are independent. Since

R. Magnus, *Essential Ordinary Differential Equations*, Springer Undergraduate Mathematics Series, https://doi.org/10.1007/978-3-031-11531-8_7

these vectors are

$$\big(y_1(x_0),\ p(x_0)y_1'(x_0)\big) \quad \text{and} \quad \big(y_2(x_0),\ p(x_0)y_2'(x_0)\big),$$

and since $p(x_0) \neq 0$, this is equivalent to the familiar requirement that the Wronskian

$$W(y_1, y_2)(x) := \begin{vmatrix} y_1(x) & y_2(x) \\ y_1'(x) & y_2'(x) \end{vmatrix}$$

does not vanish at $x = x_0$. The point x_0 is here any point in $[a, b]$ (we can admit an endpoint here). Therefore if $W(y_1, y_2)(x)$ is non-zero at one point of $[a, b]$ it is non-zero at all. In fact it is easily shown that $W(y_1, y_2)(x)$ is necessarily a constant multiple of the function $1/p(x)$, a fact that we shall frequently appeal to.

Exercise Verify the claims made in the previous paragraph.

Now let u_1 and u_2 form a solution basis and normalise them so that

$$W(u_1, u_2)(x) = 1/p(x).$$

The variation of constants formula (Proposition 1.9) now gives a nice particular solution for the equation

$$\frac{d}{dx}\Big(p(x)\frac{dy}{dx}\Big) + q(x)y = g(x),$$

namely,

$$y(x) = -u_1(x)\int u_2(x)g(x)\,dx + u_2(x)\int u_1(x)g(x)\,dx.$$

In particular the solution that satisfies $y(a) = y'(a) = 0$ is

$$y(x) = -u_1(x)\int_a^x u_2\,g + u_2(x)\int_a^x u_1\,g.$$

Exercise Show that the second order equation

$$p_2(x)y'' + p_1(x)y' + p_0(x)y = 0$$

can be cast in the Sturm-Liouville form by multiplying it by $e^{m(x)}$, where $m(x)$ is an antiderivative of $p_1(x)/p_2(x)$.

7.1 Symmetry and Self-adjointness

We consider the *differential operator*

$$Ly := \frac{d}{dx}\left(p(x)\frac{dy}{dx}\right) + q(x)y$$

for a class of functions $y(x)$ defined in the interval $[a, b]$.

Definition The operator L is called a *regular Sturm-Liouville operator* if the coefficient functions $p(x)$ and $q(x)$ are real valued and continuous in the closed and bounded interval $[a, b]$, and $p(x) \neq 0$ for $a \leq x \leq b$.[1]

Throughout the whole of this section, we maintain the assumption that the operator L is a regular Sturm-Liouville operator.

In order to make the exposition flow more smoothly we shall also assume that $p(x)$ is continuously differentiable and that p' extends to a continuous function in $[a, b]$. Then Ly certainly makes sense if $y(x)$ has second order derivatives in $]a, b[$ and is such that y, y' and y'' extend to continuous functions in $[a, b]$. The functions $y(x)$ just described constitute a space denoted by $C^2[a, b]$. In such a case Ly is continuous in the closed interval $[a, b]$, so that L may be viewed as linear mapping

$$L : C^2[a, b] \to C[a, b]$$

from one vector space of functions to another. We shall usually allow our functions to take complex values, so that the domain and codomain of L are vector spaces over the complex field $\mathbb{C}$. We have seen the advantages of allowing complex solutions, although the equation has only real coefficients. However, we never deviate from the assumption that $p(x)$ and $q(x)$ are real valued.

For complex valued functions u and v integrable in $[a, b]$ we define

$$\langle u, v\rangle := \int_a^b u(x)\overline{v(x)}\, dx$$

This is an example of a Hermitean inner product defined on the vector space of integrable functions. It has the obvious properties of being linear in u, and conjugate linear in v. It is conjugate symmetric

$$\langle v, u\rangle = \overline{\langle u, v\rangle},$$

[1] Frequently the Sturm-Liouville operator is written with a preceding minus sign. This is because the function $p(x)$ is often taken to be negative in practical applications, as the exercises in this section demonstrate. The aesthetics behind this need not concern us.

and positive semidefinite

$$\langle u, u \rangle \geq 0,$$

but we can have $\langle u, u \rangle = 0$ though u is not entirely the zero-function (though if u is also continuous then $u = 0$). This causes no problems in practice.

The mean square of a function u, integrable on $[a, b]$, is the quantity

$$\langle u, u \rangle = \int_a^b |u|^2$$

(though, properly, one should divide by $b - a$ before calling it a mean) and the L^2-norm is

$$\|u\|_2 := \sqrt{\langle u, u \rangle}.$$

We can expand the L^2-norm of a sum of functions using the algebraic properties of the inner product just detailed:

$$\left\| \sum_{j=1}^{n} u_j \right\|_2^2 = \sum_{j=1}^{n} \|u_j\|_2^2 + 2 \sum_{1 \leq j < k \leq n} \text{Re} \langle u_j, u_k \rangle.$$

In particular if $\langle u_j, u_k \rangle = 0$ for $j \neq k$, we have the *Pythagorean rule*:

$$\left\| \sum_{j=1}^{n} u_j \right\|_2^2 = \sum_{j=1}^{n} \|u_j\|_2^2.$$

Other important properties (for which we refer the reader to a text on metric spaces or algebra) are the Cauchy-Schwarz inequality

$$|\langle u, v \rangle| \leq \|u\|_2 \|v\|_2$$

and Minkowski's inequality (essentially the triangle inequality)

$$\|u + v\|_2 \leq \|u\|_2 + \|v\|_2.$$

Now if u and v are in $C^2[a,b]$ and they both have compact support[2] in $]a,b[$, we find, integrating by parts, that

$$\begin{aligned}\langle u, Lv\rangle &= \int_a^b u(x)\Big(\big(p(x)\overline{v'(x)}\big)' + q(x)\overline{v(x)}\Big)\,dx\\ &= \int_a^b \Big(\big(p(x)u'(x)\big)' + q(x)u(x)\Big)\overline{v(x)}\,dx\\ &= \langle Lu, v\rangle.\end{aligned}$$

The conclusion, that $\langle u, Lv\rangle = \langle Lu, v\rangle$, expresses *symmetry* of the differential operator L, a property that is of paramount importance for what follows. However, if we only assume that u and v are in $C^2[a,b]$, without necessarily having compact support, then the integrations produce boundary terms, spoiling the symmetry, and the previous calculation leads to

$$\begin{aligned}\langle u, Lv\rangle - \langle Lu, v\rangle &= p(b)u(b)\overline{v'(b)} - p(a)u(a)\overline{v'(a)}\\ &\qquad - p(b)u'(b)\overline{v(b)} + p(a)u'(a)\overline{v(a)}\\ &= p(b)\begin{vmatrix} u(b) & u'(b)\\ \overline{v(b)} & \overline{v'(b)}\end{vmatrix} - p(a)\begin{vmatrix} u(a) & u'(a)\\ \overline{v(a)} & \overline{v'(a)}\end{vmatrix}\end{aligned} \tag{7.3}$$

We can recover the symmetry property $\langle u, Lv\rangle = \langle Lu, v\rangle$, if u and v both satisfy the same, suitably chosen, *homogeneous boundary conditions*

$$U_1(y) = 0, \quad U_2(y) = 0,$$

that cause the boundary terms arising from the partial integrations (the right-hand side of (7.3)) to be 0. We always assume our boundary conditions to be linear functionals, which should be independent, for otherwise they would not genuinely be two conditions (recall the discussion in Sect. 1.5).

In order to verify the symmetry property for given boundary conditions, the right-hand side of (7.3) can be thought of as a function of the eight variables, $u(a)$, $u'(a)$, $v(a)$, $v'(a)$, $u(b)$, $u'(b)$, $v(b)$ and $v'(b)$, subject to four linear conditions, the boundary conditions applied to u and v. So the eight variables can be reduced to four *independent variables*.

[2] The support of a function is the closure of the set where it is non-zero. Having compact support in $]a,b[$ therefore means that there exist a' and b', such that $a < a' < b' < b$ and the function is zero for $a < x < a'$ and for $b' < x < b$. The concept will not be used beyond this paragraph, though it was briefly mentioned in 1.4 (project on Schrödinger's equation).

There are two principal cases that have practical applications, and the verification of symmetry is very easy:

A. *Separated boundary conditions*. If $u(x)$ and $v(x)$ both satisfy boundary conditions $U_1(y) = 0$ and $U_2(y) = 0$, with

$$U_1(y) = \alpha_1 y(a) + \alpha_2 y'(a), \quad U_2(y) = \beta_1 y(b) + \beta_2 y'(b),$$

where the vectors (α_1, α_2) and (β_1, β_2) are real and neither is $(0, 0)$, then the determinants in (7.3) are both zero.

B. *Periodic boundary conditions*. If $p(a) = p(b)$ and if $u(x)$ and $v(x)$ both satisfy boundary conditions $U_1(y) = 0$ and $U_2(y) = 0$, with

$$U_1(y) = y(a) - y(b), \quad U_2(y) = y'(a) - y'(b)$$

then the boundary terms rather obviously total to 0.

Definition The boundary value problem

$$Ly = h(x), \quad (a < x < b),$$
$$U_1(y) = c_1, \quad U_2(y) = c_2,$$

where L is a regular Sturm-Liouville operator, is called *self-adjoint*, if, whenever $u(x)$ and $v(x)$ are functions in $C^2[a, b]$ that satisfy $U_1(u) = U_2(u) = U_1(v) = U_2(v) = 0$, then $\langle u, Lv\rangle = \langle Lu, v\rangle$.

Self-adjointness is solely a property of the differential operator L and the boundary operators U_1 and U_2; the function $h(x)$ and the quantities c_1 and c_2 are not involved. Therefore we often avoid mentioning irrelevancies by saying that *the operator L is self-adjoint when accompanied by the boundary operators U_1 and U_2*. The most important cases by far are the separated boundary conditions, case A, and the periodic boundary conditions, case B.

Consider the non-homogeneous problem with homogeneous boundary conditions

$$Ly = h(x), \quad U_1(y) = 0, \quad U_2(y) = 0, \tag{7.4}$$

supposing it to be self-adjoint. If $y(x)$ is a solution and $\phi(x)$ satisfies

$$L\phi(x) = 0, \quad U_1(\phi) = 0, \quad U_2(\phi) = 0$$

then

$$\langle h, \phi\rangle = \langle Ly, \phi\rangle = \langle y, L\phi\rangle = 0.$$

That is, in order that a solution to (7.4) exists it is necessary that $\langle h, \phi\rangle = 0$, or, using geometrical terminology, that h is *orthogonal* to all solutions of the homogeneous problem. But it is also sufficient, as we now explain.

We view the mapping

$$L : u \mapsto (pu')' + qu$$

as a linear mapping from the vector space $C^2[a, b]$ to the vector space $C[a, b]$. As such it is surjective and its null-space is two-dimensional.

Now let E be the vector space of all functions f in $C^2[a, b]$ that satisfy $U_1(f) = U_2(f) = 0$. The boundary conditions are linearly independent, so that the infinite-dimensional space E has codimension 2 in $C^2[a, b]$. The null-space of the restriction of L to E

$$L|_E : E \to C[a, b]$$

has dimension 0, 1 or 2, since it is the intersection of E with the two-dimensional nullspace of L. Let r be the dimension of this space (it is either 0, 1 or 2). Then the codimension of $L(E)$ in $C[a, b]$ is also equal to r (recall the discussion of this point in Sect. 1.5).

Now we observe that the subspace of functions in $C[a, b]$, that are orthogonal to all the solutions of $L|_E u = 0$, has codimension r (since there are r independent solutions), and we saw previously that this subspace includes the range of $L|_E$. It therefore equals the range of $L|_E$, as we intended to demonstrate.

We can illustrate this with explicit calculations in the case of separated boundary conditions. Firstly, we know that if the homogeneous problem

$$Ly = 0, \quad \alpha_1 y(a) + \alpha_2 y'(a) = 0, \quad \beta_1 y(b) + \beta_2 y'(b) = 0$$

has no solution except $y = 0$, then the non-homogeneous problem

$$Ly = g(x), \quad \alpha_1 y(a) + \alpha_2 y'(a) = c_1, \quad \beta_1 y(b) + \beta_2 y'(b) = c_2$$

has a unique solution for all continuous g, and all c_1 and c_2.

Suppose next that the homogeneous problem has a non-identically-zero solution $\phi(x)$. There cannot be a second solution independent of ϕ, for otherwise all solutions of $Ly = 0$ would satisfy $\alpha_1 y(a) + \alpha_2 y'(a) = 0$, which would contradict the fundamental existence theorem. Therefore, as we argued above, the non-homogeneous problem

$$Ly = g(x), \quad \alpha_1 y(a) + \alpha_2 y'(a) = 0, \quad \beta_1 y(b) + \beta_2 y'(b) = 0, \tag{7.5}$$

has a solution provided g is orthogonal to ϕ.

To verify this claim by calculation we assume that $\langle g, \phi \rangle = 0$ and write a general solution of $Ly = g(x)$ in the form

$$y(x) = Au_1(x) + Bu_2(x) - u_1(x) \int_a^x u_2 g + u_2(x) \int_a^x u_1 g.$$

Here it is assumed that the basis comprising u_1 and u_2 is normalised to satisfy $W(u_1, u_2) = 1/p(x)$ (see the opening paragraphs of this chapter). We can also take $u_1 = \phi$. Then the boundary condition at $x = a$ is satisfied if and only if

$$B\big(\alpha_1 u_2(a) + \alpha_2 u_2'(a)\big) = 0$$

and since $\alpha_1 u_2(a) + \alpha_2 u_2'(a) \neq 0$, this holds if and only if $B = 0$. Since u_1 satisfies the boundary condition at $x = b$, the latter is satisfied by $y(x)$ if and only if

$$\big(\beta_1 u_2(b) + \beta_2 u_2'(b)\big) \int_a^b u_1 g = 0$$

and since $\beta_1 u_2(b) + \beta_2 u_2'(b) \neq 0$ this holds if and only if

$$\int_a^b u_1 g = 0.$$

But this is just our assumption $\langle g, \phi \rangle = 0$. One solution of the inhomogeneous problem is therefore

$$y(x) = -\phi(x) \int_a^x u_2 g + u_2(x) \int_a^x \phi g$$

where u_2 is a solution of $Ly = 0$ such that $W(\phi, u_2) = 1/p(x)$. Clearly we may add to this any multiple of $\phi(x)$.

We note that the solution just described is also the solution of $Ly = g$ that satisfies $y(a) = y'(a) = 0$. It might have been predicted that should the problem (7.5) have a solution in the case under consideration, where the homogeneous problem has a non-trivial solution, then one of its solutions would be the solution of $Ly = g$ that satisfies $y(a) = y'(a) = 0$.

Exercise Explain the claim made in the previous paragraph.

7.1.1 *Rayleigh Quotient*

Consider a regular Sturm-Liouville operator

$$Ly := (p(x)y')' + q(x)y$$

on the interval $[a, b]$, and suppose that it is self-adjoint with boundary conditions $U_1(y) = 0$, $U_2(y) = 0$. The quantity

$$R(u) := \frac{\langle Lu, u\rangle}{\langle u, u\rangle}$$

is called the Rayleigh quotient.[3] It is defined for all $u \in C^2[a, b]$ except for $u = 0$. For most purposes we compute $R(u)$ for normalised functions, that is for $\|u\|_2 = 1$, in which case

$$R(u) = \langle Lu, u\rangle.$$

The minimum of $R(u)$, taken over functions in $C^2[a, b]$ that satisfy the boundary conditions (excluding as always the zero-function), often appears in practical applications. For example, in a one-dimensional quantum mechanical system in a bounded domain, the operator L (with suitable boundary conditions making it self-adjoint) is the Hamiltonian. The minimum of $R(u)$ (taken over functions that satisfy the boundary conditions) is the ground state energy, a quantity of great physical importance.

Can we show that the minimum exists? There are really two questions here. Firstly, is the Rayleigh quotient bounded below? That is, does there exist a number K such that

$$R(u) \geq K$$

for all functions in $C^2[a, b]$ that satisfy the boundary conditions (excluding as always the zero-function)? If so we can set

$$m := \inf R(u).$$

Secondly, if $R(u)$ is bounded below, does there exist $\phi \in C^2[a, b]$, satisfying the boundary conditions and not identically zero, such that $R(\phi) = m$?

The two questions are logically distinct, but it was a common mistake of mathematicians, pre-twentieth century, and still of students today, to confuse infimum and minimum. We shall answer both questions in Sect. 7.2, but some progress will be made in the following exercises.

[3] Also known as the Rayleigh-Ritz quotient.

7.1.2 Exercises

1. For the operator $L(y) := -y''$ on the interval $[0, \pi]$, show that it is self-adjoint under each of the following sets of boundary conditions:

   (a) $y(0) = y(\pi) = 0$.
   (b) $y'(0) = y'(\pi) = 0$.
   (c) $y'(0) - y(0) = 0, \quad y'(\pi) + y(\pi) = 0$.

   Show that the Rayleigh quotient is strictly positive in cases (a) and (c), but only non-negative in case (b).
   Hint Transform the Rayleigh quotient using integration by parts.

   In the remaining exercises we consider a regular Sturm-Liouville operator

$$Ly := (p(x)y')' + q(x)y, \quad (a < x < b).$$

2. Suppose that $p(a) = p(b)$. We pose a generalisation of periodic boundary conditions,

$$\alpha_1 y(a) + \alpha_2 y'(a) = y(b), \quad \beta_1 y(a) + \beta_2 y'(a) = y'(b),$$

   with real coefficients α_1, α_2, β_1 and β_2. What condition on these coefficients is necessary and sufficient for the problem to be self-adjoint?
   Hint Use (7.3) and consider $u(a)$, $u'(a)$, $v(a)$ and $v'(a)$ as the independent variables.
3. Show that if $u \in C^2[a, b]$ and $\|u\|_2 = 1$ then

$$\langle Lu, u\rangle \geq p(b)u'(b)\overline{u(b)} - p(a)u'(a)\overline{u(a)} + Q - \int_a^b p|u'|^2$$

   where $Q = \inf q$. Deduce that if $p < 0$ and u is subject to boundary conditions, then the Rayleigh quotient $R(u)$ lies in the interval $[Q, \infty[$ in the following self-adjoint cases:

   (a) The boundary conditions are separated, that is,

$$\alpha_1 y(a) + \alpha_2 y'(a) = 0, \quad \beta_1 y(b) + \beta_2 y'(b) = 0,$$

   with $\alpha_1\alpha_2 \leq 0$ and $\beta_1\beta_2 \geq 0$.
   Note As usual, neither the vector (α_1, α_2) nor the vector (β_1, β_2) is $(0, 0)$.
   (b) $p(a) = p(b)$ and the boundary conditions are periodic.

4. Suppose that the Rayleigh quotient is bounded below and let $m = \inf R(u)$, the infimum being taken over all $u \in C^2[a, b]$ that satisfy the boundary conditions and are not identically zero. Now suppose that the infimum is attained, that is, there is a function $\phi \in C^2[a, b]$, satisfying the boundary conditions, such that

$R(\phi) = m$. Without loss of generality one may assume that $\|\phi\|_2 = 1$, so that in fact we have $\langle L\phi, \phi\rangle = m$.

(a) Show that for all $h \in C^2[a, b]$ satisfying the boundary conditions we have

$$\langle L(\phi + h), \phi + h\rangle \geq m\langle \phi + h, \phi + h\rangle$$

and deduce that

$$\operatorname{Re}\langle L\phi - m\phi, h\rangle \geq 0.$$

(b) Deduce that for all $h \in C^2[a, b]$ satisfying the boundary conditions we have

$$\langle L\phi - m\phi, h\rangle = 0.$$

(c) Deduce that

$$L\phi = m\phi.$$

Hint The obvious approach of taking $h = L\phi - m\phi$ in item (b) is not quite correct, since we cannot expect $L\phi - m\phi$ to be differentiable or to satisfy the boundary conditions. Therefore, think locally. If $L\phi - m\phi$ differs from 0 at $x_0 \in]a, b[$, then a contradiction can be obtained by choosing $\delta > 0$ sufficiently small, and taking h to be a function in $C^2[a, b]$, such that $h(x) = 0$ for $|x - x_0| > \delta$ and $h(x) > 0$ for $|x - x_0| < \delta$.

The result in Exercise 4 can be expressed by saying that if the infimum m is attained at a normalised function ϕ, then m is an *eigenvalue* of L (with the boundary conditions) and ϕ a corresponding *eigenfunction*. These are concepts that we define and study in the succeeding sections, and they constitute the heart of Sturm-Liouville theory. The question remains, is the minimum attained? In the next section we approach this question by studying eigenvalues.

7.2 Eigenvalues and Eigenfunctions

We consider the Sturm-Liouville operator

$$Ly := \frac{d}{dx}\left(p(x)\frac{dy}{dx}\right) + q(x)y$$

in the interval $[a, b]$, together with boundary operators U_1 and U_2 which, and this is most important for what follows, *define a self-adjoint problem*. We repeat that the functions $p(x)$ and $q(x)$ are real valued and continuous in the closed interval $[a, b]$, and that $p(x) \neq 0$ for $a \leq x \leq b$ (in short, the operator is regular). However, we shall allow the function $y(x)$, on which L and the boundary operators U_1 and U_2 act, to be complex valued.

The spaces $C[a, b]$ and $C^2[a, b]$ are intended to be spaces of complex valued functions. We therefore explicitly admit complex valued solutions. For all u and v in $C[a, b]$ we have the inner product (introduced previously for the larger space of integrable functions)

$$\langle u, v \rangle := \int_a^b u\overline{v}$$

and for all u and v in $C^2[a, b]$ that satisfy $U_1(u) = U_2(u) = U_1(v) = U_2(v) = 0$ we have the symmetry property

$$\langle Lu, v \rangle = \langle u, Lv \rangle.$$

Recall that if f is (Riemann) integrable in $[a, b]$ its L^2-norm is given by

$$\|f\|_2 = \langle f, f \rangle^{1/2}.$$

The vector space of functions integrable in $[a, b]$ includes all continuous functions. Because we are using here the Riemann integral we are assuming that the function f is bounded.[4]

Definition A complex number λ is called an eigenvalue for the operator L with the boundary operators U_1 and U_2, if the problem

$$Ly = \lambda y, \quad U_1(y) = U_2(y) = 0$$

has a non-trivial solution (that is, a solution that is not identically zero).

We immediately note that if λ is not an eigenvalue, then the problem

$$Ly - \lambda y = h(x), \quad U_1(y) = c_1, \quad U_2(y) = c_2$$

has a unique solution for all $h \in C[a, b]$ and all c_1 and c_2.

Definition If λ is an eigenvalue for the operator L with the boundary operators U_1 and U_2, then a non-trivial solution ϕ of

$$Ly = \lambda y, \quad U_1(y) = U_2(y) = 0$$

[4] One could (and should) extend all of this to the Lebesgue integrable functions; these are not necessarily bounded.

is called an eigenfunction; or, should we wish to mention λ, an eigenfunction belonging to λ.

Proposition 7.1

1. *All eigenvalues are real.*
2. *If ϕ and ψ are eigenfunctions belonging to distinct eigenvalues, then $\langle \phi, \psi \rangle = 0$.*
3. *For each eigenvalue λ, there exist at most two linearly independent eigenfunctions.*

Proof
1. Suppose that $\phi \in C^2[a, b]$ is not identically 0 and satisfies $L\phi = \lambda\phi$, $U_1(\phi) = U_2(\phi) = 0$. Then

$$\lambda\langle \phi, \phi \rangle = \langle \lambda\phi, \phi \rangle = \langle L\phi, \phi \rangle = \langle \phi, L\phi \rangle = \langle \phi, \lambda\phi \rangle = \overline{\lambda}\langle \phi, \phi \rangle$$

so that, in view of $\langle \phi, \phi \rangle \neq 0$, we must have $\lambda = \overline{\lambda}$.
2. Let λ_1 and λ_2 be distinct eigenvalues, and let ϕ_1 and ϕ_2 be eigenfunctions belonging to λ_1 and λ_2 respectively. Then, since λ_2 is real we have

$$\begin{aligned}
(\lambda_1 - \lambda_2)\langle \phi_1, \phi_2 \rangle &= \lambda_1\langle \phi_1, \phi_2 \rangle - \overline{\lambda}_2\langle \phi_1, \phi_2 \rangle \\
&= \langle \lambda_1\phi_1, \phi_2 \rangle - \langle \phi_1, \lambda_2\phi_2 \rangle \\
&= \langle L\phi_1, \phi_2 \rangle - \langle \phi_1, L\phi_2 \rangle \\
&= 0.
\end{aligned}$$

Since $\lambda_1 \neq \lambda_2$ we conclude that $\langle \phi_1, \phi_2 \rangle = 0$.
3. Obvious, since the space of solutions of $Ly - \lambda y = 0$ is two-dimensional. □

At this point there is no reason to suppose that eigenvalues exist. In fact they do, and are plentiful in a very precise sense. The main tool for proving this is the Green's function. The theoretical machinery underlying existence of eigenvalues will be complex analysis. Other treatments, going beyond what we intend to do here, rely on functional analysis, more precisely Ascoli's theorem and the rather technical theory of self-adjoint linear operators in Hilbert spaces. A treatment using complex analysis seems more elementary, goes less far, but it is likely that the reader of this text has some acquaintance with the required material. It is also historically the first approach that led to rigorous proofs of the Sturm-Liouville development (or eigenfunction expansion) of a function, at the very start of the twentieth century.[5]

[5] The classical book 'Ordinary Differential Equations' by E. L. Ince, published in 1926, credits A. Kneser, 1904, for the Sturm-Liouville development of an arbitrary function, using methods probably similar to ours. The existence of eigenvalues goes back to a famous paper by C. Sturm from 1836.

For each eigenvalue λ we call the space of solutions of

$$Ly - \lambda y = 0, \quad U_1(y) = U_2(y) = 0,$$

the eigenspace belonging to λ. It is either one-dimensional or two-dimensional. Its non-zero elements are the eigenfunctions.

For each eigenvalue λ we can select an eigenfunction ϕ, that spans the eigenspace if the latter is one-dimensional, or two eigenfunctions ϕ and ψ, that span the eigenspace if the latter is two-dimensional. We normalise the eigenfunctions so that $\|\phi\|_2 = 1$, or in the two dimensional case so that $\|\phi\|_2 = \|\psi\|_2 = 1$ and $\langle\phi, \psi\rangle = 0$. This process leads to a *maximal orthonormal set of eigenfunctions*, that is, a set S of eigenfunctions, such that for all $\phi \in S$ we have $\|\phi\|_2 = 1$, for all ϕ and ψ in S such that $\phi \neq \psi$ we have $\langle\phi, \psi\rangle = 0$; and such that the set S is maximal with these properties.

As a first step towards proving the existence of eigenvalues (and the corresponding eigenfunctions) we introduce the Green's function $G(x, \xi, \lambda)$ for the problem

$$Ly - \lambda y = g(x), \quad U_1(y) = U_2(y) = 0.$$

Here, λ is a complex number, considered variable. The Green's function only exists provided λ is not an eigenvalue. How the Green's function depends on λ is the key to the ensuing analysis. We remind the reader that the problem is supposed to be self-adjoint. Since we assume that the operator L has real coefficient functions, the Green's function $G(x, \xi, \lambda)$ is real when λ is real.

Proposition 7.2

1. *$G(x, \xi, \lambda)$ is a meromorphic function of the complex number λ.*
2. *The eigenvalues are the singularities of $G(x, \xi, \lambda)$ and they are simple poles.*
3. *Let $f \in C[a, b]$ be such that $\int_a^b G(x, \xi, \lambda) f(\xi)\, d\xi$ is an entire analytic function of λ. Then $f = 0$.*

Proof

1. Choose a point x_0 in $]a, b[$ and let $u_1(x, \lambda)$ and $u_2(x, \lambda)$ be solutions of $Ly - \lambda y = 0$ satisfying the initial conditions $y(x_0) = 1$, $y'(x_0) = 0$ and $y(x_0) = 0$, $y'(x_0) = 1$, respectively. By the discussion in Sect. 6.2.3, the solutions $u_1(x, \lambda)$ and $u_2(x, \lambda)$ are entire analytic functions of λ. By the proof of Proposition 1.24, the Green's function can be expressed in the form

$$G(x, \xi, \lambda) = \begin{cases} A_1(\xi, \lambda)u_1(x, \lambda) + A_2(\xi, \lambda)u_2(x, \lambda), & (a < x < \xi) \\ B_1(\xi, \lambda)u_1(x, \lambda) + B_2(\xi, \lambda)u_2(x, \lambda), & (\xi < x < b). \end{cases}$$

In determining the coefficient functions A_1, A_2, B_1 and B_2, the only divisions needed are, firstly, by the non-zero quantity $p(\xi)$, secondly, by the Wronskian of u_1 and u_2,

$$\begin{vmatrix} u_1 & u_2 \\ u_1' & u_2' \end{vmatrix}$$

necessarily non-zero for all λ because the solutions are linearly independent, and thirdly by the determinant

$$\begin{vmatrix} U_1(u_1) & U_2(u_1) \\ U_1(u_2) & U_2(u_2) \end{vmatrix}.$$

The second and third quantities are entire complex-analytic functions of λ. The third cannot be identically zero; its zeros are the eigenvalues, and, since we know that a non-real λ is not an eigenvalue, the determinant is non-zero if λ is not real. As it is analytic, its zeros, if any, are real, isolated and have finite multiplicity. It follows that A_1, A_2, B_1 and B_2 are meromorphic functions of λ.

This tells us that if λ_0 is a singularity of G (or equivalently, λ_0 is a zero of the determinant formed above from the boundary operators), then we have a Laurent expansion

$$G(x, \xi, \lambda) = \sum_{k=-m}^{\infty} g_k(x, \xi)(\lambda - \lambda_0)^k$$

valid for $0 < |\lambda - \lambda_0| < r$ (for some $r > 0$ and integer m). The coefficients are given by the Cauchy integral formula

$$g_k(x, \xi) = \frac{1}{2\pi i} \int_C G(x, \xi, \lambda)(\lambda - \lambda_0)^{-k-1}\, d\lambda$$

where the contour is a positively oriented circle surrounding the singularity λ_0 and enclosing no other singularity.

2. Let λ_0 be a singularity of G. We shall show that in the Laurent expansion just described, if $m \geq 2$ then $g_{-m} = 0$.

Let $f \in C[a, b]$ and let

$$F(x, \lambda) = \int_a^b G(x, \xi, \lambda) f(\xi)\, d\xi.$$

There exists $r > 0$, such that if $0 < |\lambda - \lambda_0| < r$, then λ is not an eigenvalue, in which case $F(x, \lambda)$ is a C^2 function of x, satisfies the boundary conditions, and

$$LF(x, \lambda) = \lambda F(x, \lambda) + f(x),$$

so that,

$$(L - \lambda_0)F(x, \lambda) = (\lambda - \lambda_0)F(x, \lambda) + f(x). \tag{7.6}$$

It is important to observe that F, and its partial derivatives $\partial F/\partial x$ and $\partial^2 F/\partial x^2$, are actually continuous functions of the pair (x, λ), for $a \le x \le b$ and $0 < |\lambda - \lambda_0| < r$. We leave it to the reader to provide the details, but it boils down to observing first that the same is true of the partial derivatives with respect to x of u_1 and u_2 up to order 2 (as observed at the very end of Sect. 6.2), and studying closely the structure of the Green's function.

The function $F(x, \lambda)$ has a Laurent expansion

$$F(x, \lambda) = \sum_{k=-m}^{\infty} c_k(x)(\lambda - \lambda_0)^k, \quad (0 < |\lambda - \lambda_0| < r)$$

where, using a contour C consisting of a positively oriented circle centred at λ_0 and with radius less than r, we have

$$\begin{aligned} c_k(x) &= \frac{1}{2\pi i}\int_C F(x, \lambda)(\lambda - \lambda_0)^{-k-1}\, d\lambda \\ &= \frac{1}{2\pi i}\int_C \left(\int_a^b G(x, \xi, \lambda) f(\xi)\, d\xi\right)(\lambda - \lambda_0)^{-k-1}\, d\lambda \\ &= \int_a^b g_k(x, \xi) f(\xi)\, d\xi \end{aligned}$$

the last equality being an application of Fubini's theorem (reversing the order of the integrals). A simple argument (left to the reader) shows that we can conclude that $g_k = 0$ if we can show that $c_k = 0$ whatever the choice of $f \in C[a, b]$.

A further point, indicated by the first line of the above display, is that each coefficient function c_k satisfies the homogeneous boundary conditions (because $F(x, \lambda)$ does so), and, the observation of the paragraph before last justifying differentiation across the integral sign, that each coefficient function c_k is in $C^2[a, b]$.

Using (7.6) we can write, for $0 < |\lambda - \lambda_0| < r$ and $a \le x \le b$:

$$(L - \lambda_0)\sum_{k=-m}^{\infty} c_k(x)(\lambda - \lambda_0)^k = \sum_{k=-m}^{\infty} c_k(x)(\lambda - \lambda_0)^{k+1} + f(x) \tag{7.7}$$

from which we can deduce

$$(L-\lambda_0)c_{-m}(x)=0$$
$$(L-\lambda_0)c_k(x)=c_{k-1}(x),$$
$$(k=-m+1,-m+2,\ldots \text{ and so on, but } k\neq 0)$$
$$(L-\lambda_0)c_0(x)=c_{-1}(x)+f(x). \tag{7.8}$$

The deduction of (7.8) from (7.7) is plausible, being based on carrying the operator $L-\lambda_0$ across the summation sign. Given that this is permissible, the reader is invited to finish the deduction. In order not to hold up the flow, we shall consider why the passage of $L-\lambda_0$ across the summation sign is permissible after the body of the proof. It requires a little more complex analysis and the reader might skip it.

We claim that if $m\geq 2$ then $c_{-m}=0$, so that F has, at worst, a simple pole at λ_0. This follows by noting that if $m\neq 1$ then $c_{-m}=(L-\lambda_0)c_{-m+1}$ and $(L-\lambda_0)c_{-m}=0$ so that

$$\langle c_{-m},c_{-m}\rangle=\big\langle (L-\lambda_0)c_{-m+1},c_{-m}\big\rangle=\big\langle c_{-m+1},(L-\lambda_0)c_{-m}\big\rangle=0$$

whence $c_{-m}=0$.

From the fact that F has at worst a simple pole at λ_0 for each $f\in C[a,b]$, we conclude, by reasoning already given, that the same is true for G. Since λ_0 is a singularity of G it must be an actual simple pole.

3. Let $F(x,\lambda)=\int_a^b G(x,\xi,\lambda)f(\xi)\,d\xi$. We have a power series expansion

$$F(x,\lambda)=\sum_{k=0}^{\infty}c_k(x)\lambda^k$$

valid for all complex λ and $a\leq x\leq b$. There exists $r>0$, such that for $0<|\lambda|<r$ we have

$$LF(x,\lambda)=\lambda F(x,\lambda)+f(x),\quad U_1\big(F(x,\lambda)\big)=U_2\big(F(x,\lambda)\big)=0.$$

Moreover, by the same reasoning as in part 2, the coefficient functions c_k are in $C^2[a,b]$, satisfy the homogeneous boundary conditions, and the equations

$$Lc_k(x)=c_{k-1}(x)\quad (k\neq 0),\qquad Lc_0(x)=f(x).$$

We note that for $j\geq 1$ and $k\geq 1$ we have

$$\langle c_{j-1},c_k\rangle=\langle Lc_j,c_k\rangle=\langle c_j,Lc_k\rangle=\langle c_j,c_{k-1}\rangle.$$

It follows that the inner product $\langle c_j, c_k \rangle$ depends only on the sum $j + k$. Defining the sequence $w_n := \langle c_0, c_n \rangle$, we have

$$\langle c_j, c_k \rangle = \langle c_0, c_{j+k} \rangle = w_{j+k}.$$

Note that $w_{2n} = \langle c_n, c_n \rangle$ and is therefore non-negative (though it could be 0). The function

$$\langle F, c_0 \rangle = \sum_{n=0}^{\infty} w_n \lambda^n \tag{7.9}$$

is an entire function of λ (the reader is invited to ponder on the plausible equality sign here). However, for $n = 1, 2, ...$ we have, by the Cauchy-Schwarz inequality

$$w_{2n}^2 = \langle c_{n-1}, c_{n+1} \rangle^2 \leq \langle c_{n-1}, c_{n-1} \rangle \langle c_{n+1}, c_{n+1} \rangle = w_{2n-2} w_{2n+2}$$

so that, if we assume that w_2 is non-zero, we can deduce that all coefficients w_{2n} are non-zero, and

$$\frac{w_{2n}}{w_{2n-2}} \leq \frac{w_{2n+2}}{w_{2n}}$$

that is, the ratio w_{2n}/w_{2n-2} is increasing. But this means that the series

$$\sum_{n=0}^{\infty} w_{2n} \lambda^{2n}$$

cannot have infinite radius of convergence, whereas, on the contrary, it must converge for all λ just as the series $\sum_{n=0}^{\infty} w_n \lambda^n$ does.

The assumption that $w_2 \neq 0$ therefore cannot hold and we conclude that $w_2 = 0$. Since $w_2 = \|c_1\|_2^2$ we must have $c_1 = 0$. Hence $c_0 = Lc_1 = 0$ and $f = Lc_0 = 0$. □

Why we may pass $L - \lambda_0$ across the summation sign in (7.7)

An explicit form for the remainder in the Laurent expansion is available (the reader should consult a text on complex analysis), and gives

$$F(x, \lambda) = \sum_{k=-m}^{n-1} c_k(x)(\lambda - \lambda_0)^k + R_n(x, \lambda)$$

where, n is a positive integer and

$$R_n(x, \lambda) = \frac{1}{2\pi i} \int_C \left(\frac{\lambda - \lambda_0}{\zeta - \lambda_0} \right)^n \frac{F(x, \zeta)}{\zeta - \lambda} \, d\zeta.$$

The contour is a circle $|\zeta - \lambda_0| = \rho$, with $\rho < r$, enclosing λ. We wish to show that

$$\lim_{n\to\infty} (L - \lambda_0)R_n(x, \lambda) = 0.$$

Since $F(x, \lambda)$ is a C^2 function of x (as observed previously) we may pass $L - \lambda_0$ across the integral sign, and, using (7.6), write

$$\begin{aligned}(L - \lambda_0)R_n(x, \lambda) &= \frac{1}{2\pi i}\int_C \left(\frac{\lambda - \lambda_0}{\zeta - \lambda_0}\right)^n \frac{(\zeta - \lambda_0)F(x, \zeta) + f(x)}{\zeta - \lambda}\, d\zeta \\ &= \frac{1}{2\pi i}\int_C \left(\frac{\lambda - \lambda_0}{\zeta - \lambda_0}\right)^n \frac{F(x, \zeta)}{1 - \frac{\lambda-\lambda_0}{\zeta-\lambda_0}}\, d\zeta\,.\end{aligned}$$

It is an exercise in analysis, left to the reader, to show that this converges to 0 as $n \to \infty$. A similar use of the remainder term justifies the equality in (7.9).

Proposition 7.3

1. *The set of eigenvalues is a closed, discrete, non-empty subset of the real line* $\mathbb{R}$.
2. *Let* $(\phi_k)_{k=1}^{\infty}$, *or* $(\phi_k)_{k=1}^{N}$ *if finite, be a maximal orthonormal sequence of eigenfunctions.*[6] *Let* $f \in C[a, b]$ *be such that* $\langle \phi_k, f\rangle = 0$ *for all* k. *Then* $f = 0$.

Proof

1. The eigenvalues are the zeros of an entire analytic function, hence form a discrete closed set. If this set is empty then the function $\int_a^b G(x, \xi, \lambda)\, d\xi$ is an entire analytic function of λ. This contradicts the previous proposition (part 3).
2. We shall show that the function $\int_a^b G(x, \xi, \lambda) f(\xi)\, d\xi$ is an entire analytic function of λ, from which $f = 0$ follows by the previous proposition, part 3.

It is enough to show that if λ_0 is an eigenvalue, $f \in C[a, b]$ and f is orthogonal to all eigenfunctions belonging to λ_0, then $\int_a^b G(x, \xi, \lambda) f(\xi)\, d\xi$ has a removable singularity at λ_0. In fact, for $f \in C[a, b]$, we have, by the previous proposition, part 2, that

$$\int_a^b G(x, \xi, \lambda) f(\xi)\, d\xi = \sum_{k=-1}^{\infty} c_k(x)(\lambda - \lambda_0)^k$$

for λ near to λ_0. In the proof of the previous proposition, part 2, it appeared that

$$(L - \lambda_0)c_{-1}(x) = 0, \quad (L - \lambda_0)c_0(x) = c_{-1}(x) + f(x)$$

[6] For all we know at this stage the sequence could be finite.

We want to show that $c_{-1} = 0$. Now if $c_{-1} \neq 0$, then it is an eigenfunction belonging to λ_0, and therefore orthogonal to f. Therefore

$$0 = \langle f, c_{-1} \rangle = \big\langle (L - \lambda_0)c_0, c_{-1} \big\rangle - \langle c_{-1}, c_{-1} \rangle$$
$$= \big\langle c_0, (L - \lambda_0)c_{-1} \big\rangle - \langle c_{-1}, c_{-1} \rangle = -\langle c_{-1}, c_{-1} \rangle.$$

We deduce that $\langle c_{-1}, c_{-1} \rangle = 0$, that is, $c_{-1} = 0$, a straightforward contradiction. We conclude that $c_{-1} = 0$. □

Proposition 7.4 *The set of eigenvalues is infinite and can be enumerated as a sequence* $(\lambda_n)_{n=0}^{\infty}$ *such that* $\lim_{n\to\infty} |\lambda_n| = \infty$.

Proof Because the set of eigenvalues is closed and discrete it has no limit point. It is therefore enough to show that it is infinite. If it is finite then there is a finite maximal orthonormal set of eigenfunctions, which we can enumerate as a sequence $(\phi_n)_{n=1}^{N}$. The vector space $C[a, b]$ is infinite-dimensional. Therefore there exists $f \in C[a, b]$, such that f is not in the span of the functions ϕ_n, $(n = 1, ..., N)$. But now the function

$$g := f - \sum_{n=1}^{N} \langle f, \phi_n \rangle \phi_n$$

satisfies $\langle g, \phi_n \rangle = 0$ for each n. By Proposition 7.3, part 2, we have $g = 0$, which contradicts the choice of f. □

7.2.1 Eigenfunction Expansions

Definition An *orthonormal sequence of functions* integrable on $[a, b]$ is a sequence, either finite $(\psi_n)_{n=1}^{N}$, or infinite $(\psi_n)_{n=1}^{\infty}$, comprising functions integrable on $[a, b]$, and such that $\|\psi_n\|_2 = 1$ for all n, and $\langle \psi_j, \psi_k \rangle = 0$ if $j \neq k$.

We present a general property of orthonormal sequences of functions:

Proposition 7.5 (Bessel's Inequality) *Let* $(\phi_n)_{n=1}^{\infty}$ *be an orthonormal sequence of functions integrable on the interval* $[a, b]$. *Let* f *be integrable on* $[a, b]$. *Then*

$$\sum_{n=1}^{\infty} |\langle f, \phi_n \rangle|^2 \leq \|f\|_2^2.$$

Proof By the algebraic properties of the inner product, for each integer N we have:

$$0 \leq \|f - \sum_{n=1}^{N} \langle f, \phi_n \rangle \phi_n\|_2^2 = \|f\|_2^2 - \sum_{n=1}^{N} |\langle f, \phi_n \rangle|^2.$$

Let N tend to infinity. □

Next we obtain a first result on the expansion of a function in a series of eigenfunctions. We assume a self-adjoint problem with Sturm-Liouville operator $Ly := -(py')' + qy$ on an interval $[a, b]$ and boundary operators U_1 and U_2. We repeat the standing assumptions for this section, that $p(x)$ and $q(x)$ are continuous in the closed and bounded interval $[a, b]$, and that $p(x) \neq 0$ for $a \leq x \leq b$. We let $(\phi_n)_{n=1}^{\infty}$ be an enumeration of a maximal orthonormal set of eigenfunctions.

Proposition 7.6 *Let $f \in C^2[a, b]$ and suppose that $U_1(f) = U_2(f) = 0$. Then*

$$f = \sum_{n=1}^{\infty} \langle f, \phi_n \rangle \phi_n$$

where the convergence is uniform in the interval $[a, b]$.

Proof We may assume that 0 is not an eigenvalue, for otherwise we can replace L by $L + \mu$ for a suitable μ, which does not change the eigenfunctions.

With this assumption we let $L\phi_n = \lambda_n \phi_n$ for $n = 1, 2, \ldots$. For clarity we note that in the sequence $(\lambda_n)_{n=1}^{\infty}$ the same eigenvalue may appear twice, if corresponding to it are two linearly independent eigenfunctions. Let $G(x, \xi)$ be the Green's function for L and the given boundary operators. Then

$$\lambda_n^{-1} \phi_n(x) = \int_a^b G(x, \xi) \phi_n(\xi)\, d\xi, \quad (n = 1, 2, \ldots)$$

We first prove that the series $\sum_{k=1}^{\infty} \langle f, \phi_k \rangle \phi_k(x)$ is uniformly convergent. We have

$$\langle f, \phi_n \rangle = \lambda_n^{-1} \langle f, L\phi_n \rangle = \lambda_n^{-1} \langle Lf, \phi_n \rangle.$$

Hence

$$\sum_{k=m}^{n} \langle f, \phi_k \rangle \phi_k(x) = \sum_{k=m}^{n} \langle Lf, \phi_k \rangle \lambda_k^{-1} \phi_k(x) = \int_a^b G(x, \xi) \sum_{k=m}^{n} \langle Lf, \phi_k \rangle \phi_k(\xi)\, d\xi$$

By the Cauchy-Schwarz inequality and the Pythagorian rule:

$$\Big|\sum_{k=m}^{n}\langle f,\phi_k\rangle\phi_k(x)\Big| \le \Big(\int_a^b |G(x,\xi)|^2\,d\xi\Big)^{1/2}\Big\|\sum_{k=m}^{n}\langle Lf,\phi_k\rangle\phi_k\Big\|_2$$
$$= \Big(\int_a^b |G(x,\xi)|^2\,d\xi\Big)^{1/2}\Big(\sum_{k=m}^{n}|\langle Lf,\phi_k\rangle|^2\Big)^{1/2}$$
$$\le M\Big(\sum_{k=m}^{n}|\langle Lf,\phi_k\rangle|^2\Big)^{1/2}$$

where, in the last line, M is a constant independent of x in the interval $[a, b]$. To be precise

$$M = \sup_{a\le x\le b}\Big(\int_a^b |G(x,\xi)|^2\,d\xi\Big)^{1/2}.$$

By Bessel's inequality (Proposition 7.5) the numerical series $\sum_{k=1}^{\infty}|\langle Lf,\phi_k\rangle|^2$ is convergent. Hence the series $\sum_{k=1}^{\infty}\langle f,\phi_k\rangle\phi_k(x)$ is uniformly convergent.

Now set

$$g(x) = \sum_{k=1}^{\infty}\langle f,\phi_k\rangle\phi_k(x), \quad (a \le x \le b).$$

The function g is continuous. Since uniform convergence allows term-by-term integration we see that

$$\langle g,\phi_n\rangle = \Big\langle \sum_{k=1}^{\infty}\langle f,\phi_k\rangle\phi_k,\phi_n\Big\rangle = \sum_{k=1}^{\infty}\langle f,\phi_k\rangle\langle\phi_k,\phi_n\rangle = \langle f,\phi_n\rangle$$

that is, $\langle f-g,\phi_n\rangle = 0$ for all n. By Proposition 7.3 we conclude that $f = g$. □

It can be useful to note that the hypotheses of Proposition 7.6 are stronger than needed to obtain the conclusion. As far as f is concerned we require two things, in addition to satisfying the boundary conditions: firstly, that Lf is integrable, so that Bessel's inequality is valid for it; and secondly, the truth of $\langle f, L\phi_n\rangle = \langle Lf,\phi_n\rangle$, which is a consequence of integration by parts. For both, it is enough if f' is piecewise continuously differentiable (instead of the stronger assumption that f is in $C^2[a, b]$).

We can use Proposition 7.6 to solve the non-homogeneous problem.

Proposition 7.7 *Consider the problem*

$$Ly - \lambda y = h(x), \quad U_1(y) = U_2(y) = 0,$$

assumed to be self-adjoint, and let $h \in C[a, b]$.

1. *If λ is not an eigenvalue then the unique solution is given by*

$$y(x) = \sum_{n=1}^{\infty} (\lambda_n - \lambda)^{-1} \langle h, \phi_n \rangle \phi_n(x).$$

2. *If λ is an eigenvalue then a solution exists if and only if $\langle h, \phi_n \rangle = 0$ for all n such that $\lambda_n = \lambda$ (there are at most two such values for n). One such solution is given by*

$$y(x) = \sum_{\lambda_n \neq \lambda} (\lambda_n - \lambda)^{-1} \langle h, \phi_n \rangle \phi_n(x).$$

In both cases the series is uniformly convergent.

Proof We have already seen that a unique solution exists in case 1, and a solution exists in case 2 if and only if h is orthogonal to all ϕ_n for which $\lambda_n = \lambda$. The solution is in $C^2[a, b]$ and satisfies the homogeneous boundary conditions, so by Proposition 7.6 it satisfies

$$y(x) = \sum_{n=1}^{\infty} \langle y, \phi_n \rangle \phi_n(x)$$

with uniform convergence of the series. But for $\lambda_n \neq \lambda$ we have

$$\begin{aligned}\langle y, \phi_n \rangle &= (\lambda_n - \lambda)^{-1} \langle y, (L - \lambda)\phi_n \rangle \\ &= (\lambda_n - \lambda)^{-1} \langle (L - \lambda) y, \phi_n \rangle \\ &= (\lambda_n - \lambda)^{-1} \langle h, \phi_n \rangle.\end{aligned}$$

□

7.2.2 *Mean Square Convergence of Eigenfunction Expansions*

If the function f is integrable we can form the function series

$$\sum_{n=1}^{\infty} \langle f, \phi_n \rangle \phi_n,$$

but we cannot expect it to converge uniformly to f, since, for example, the latter is not necessarily continuous. However, a weaker convergence notion is available, which the reader may have studied in connection with metric space theory.

Definition A function series $\sum_{n=1}^{\infty} g_n$, comprising integrable functions, is convergent *in the mean square* and its sum is the integrable function h if

$$\lim_{N\to\infty} \left\| h - \sum_{n=1}^{N} g_n \right\|_2 = 0.$$

We write

$$\sum_{n=1}^{\infty} g_n = h,$$

appending "in the mean square", if necessary, to distinguish the convergence from other kinds, such as point-wise or uniform convergence.

The formula

$$\left\| f - \sum_{n=1}^{N} \langle f, \phi_n \rangle \phi_n \right\|_2^2 = \| f \|_2^2 - \sum_{n=1}^{N} |\langle f, \phi_n \rangle|^2,$$

which appeared in the proof of Bessel's inequality, shows that the formula

$$\sum_{n=1}^{\infty} \langle f, \phi_n \rangle \phi_n = f \tag{7.10}$$

holds in the mean square if and only if Bessel's inequality is actually an equality; that is, if and only if

$$\sum_{n=1}^{\infty} |\langle f, \phi_n \rangle|^2 = \| f \|_2^2.$$

This equality is called Parseval's equation. But the formula tells us more: the mean square error

$$\left\| f - \sum_{n=1}^{N} \langle f, \phi_n \rangle \phi_n \right\|_2$$

actually decreases with increasing N. Hence to prove that the formula (7.10) holds in the mean square, it is enough to show that for each $\varepsilon > 0$, there exists N, such that

$$\left\| f - \sum_{n=1}^{N} \langle f, \phi_n \rangle \phi_n \right\|_2 < \varepsilon.$$

We don't have to consider the sum with more than N terms. This observation can be combined with the following result:

Proposition 7.8 *Let $(\phi_n)_{n=1}^N$ be an orthonormal sequence of N functions integrable on the interval $[a, b]$. Let f be integrable on the interval $[a, b]$. Then the quantity*

$$\left\| f - \sum_{n=1}^{N} d_n \phi_n \right\|_2,$$

depending on the arbitrary complex numbers $d_1, ..., d_N$, is minimised by the choice $d_n = \langle f, \phi_n \rangle$, $n = 1, ..., N$.

Proof The conclusion follows from the formula

$$\left\| f - \sum_{n=1}^{N} d_n \phi_n \right\|_2^2 = \| f \|_2^2 - \sum_{n=1}^{N} |\langle f, \phi_n \rangle|^2 + \sum_{n=1}^{N} |d_n - \langle f, \phi_n \rangle|^2$$

which is easily obtained by algebraic manipulation, left to the reader. □

Proposition 7.9 *Let $(\phi_n)_{n=1}^\infty$ be an orthonormal sequence of functions integrable on the interval $[a, b]$. Suppose that the function f, integrable in $[a, b]$ can be approximated with arbitrary accuracy in the mean square by linear combinations of the functions $\{\phi_n : n = 1, 2, ...\}$. Then*

$$f = \sum_{n=1}^{\infty} \langle f, \phi_n \rangle \phi_n$$

in the mean square.

Proof Let $\varepsilon > 0$. The stated assumption about approximating f in the mean square means that there exists a finite set $S \subset \mathbb{N}$, and complex numbers $(d_j)_{j \in S}$, such that

$$\left\| f - \sum_{j \in S} d_j \phi_j \right\|_2 < \varepsilon.$$

By Proposition 7.8 this implies that if N is so large that $S \subset \{1, ..., N\}$ then

$$\left\| f - \sum_{j=1}^{N} \langle f, \phi_j \rangle \phi_j \right\|_2 < \varepsilon.$$

As we have seen, we can conclude from this that

$$f = \sum_{n=1}^{\infty} \langle f, \phi_n \rangle \phi_n$$

in the mean square. □

The following concept is really a notion drawn from the theory of Hilbert spaces, which the reader may have encountered in a course on functional analysis.

Definition The orthonormal sequence $(\phi_n)_{n=1}^{\infty}$ is said to be *complete* if

$$f = \sum_{n=1}^{\infty} \langle f, \phi_n \rangle \phi_n$$

in the mean square, for every integrable function f.

Completeness of an orthonormal sequence is clearly a most desirable situation and we immediately apply the preceding considerations to the eigenfunctions of a Sturm-Liouville operator L.

Proposition 7.10 *Let the regular Sturm-Liouville operator L with boundary operators U_1 and U_2 define a self-adjoint problem on the interval $[a, b]$. Let $(\phi_n)_{n=1}^{\infty}$ be a maximal orthonormal sequence of eigenfunctions. Then for each function f, integrable on $[a, b]$, we have:*

1. $f = \sum_{n=1}^{\infty} \langle f, \phi_n \rangle \phi_n$ *in the mean square.*
2. $\int_a^b |f|^2 = \sum_{n=1}^{\infty} |\langle f, \phi_n \rangle|^2.$

Moreover, if f and g are integrable functions then

3. $\langle f, g \rangle = \sum_{n=1}^{\infty} \langle f, \phi_n \rangle \langle \phi_n, g \rangle$

The first conclusion says that the orthonormal sequence $(\phi_n)_{n=1}^{\infty}$ is complete.

Proof

1. It is enough to show that a function f, integrable on $[a, b]$, can be approximated arbitrarily closely in the mean square by linear combinations of the functions ϕ_n, $(n = 1, 2, ...)$.

Let f be integrable on $[a, b]$. We mean here the Riemann-Darboux integral, so that f is supposed to be bounded. We may also suppose that f is real. We introduce the bound $M = \sup |f|$. We are going to approximate f in the mean square by a function in $C^2[a, b]$ that satisfies the homogeneous boundary conditions.

Let $\varepsilon > 0$. By definition of the integral (for example using Riemann sums) we may approximate f *in the mean* by step functions; precisely, there exists a step function $s(x)$, such that

$$\int_a^b |f - s| < \varepsilon^2/18M.$$

We may suppose that $\sup |s| \leq M$, and we may replace $s(x)$ by zero in neighbourhoods $[a, a + h]$ and $[b - h, b]$ of the endpoints (for some suitably small h; how small depends only on ε and M) without affecting the validity of the above inequality. We then have

$$\int_a^b |f - s|^2 < 2M \int_a^b |f - s| < \varepsilon^2/9$$

so that

$$\|f - s\|_2 < \varepsilon/3.$$

Next we approximate s in the mean square by a C^2 function g, such that

$$\|s - g\|_2 < \varepsilon/3$$

while ensuring that g vanishes in the intervals $[a + \frac{1}{2}h]$ and $[b - \frac{1}{2}h, b]$ (see Fig. 7.1[7] showing how to approximate s *in the mean*; it is thence a short step to a mean square approximation). Then the function g satisfies $U_1(g) = U_2(g) = 0$ and

$$\|f - g\|_2 < \frac{\varepsilon}{3} + \frac{\varepsilon}{3} = \frac{2\varepsilon}{3}.$$

[7] To quote J. E. Littlewood (in 'Littlewood's miscellany'): "A heavy warning used to be given that pictures are not rigorous; this has never had its bluff called and has permanently frightened its victims into playing for safety. Some pictures, of course, are not rigorous, but I should say most are (and I use them whenever possible myself)."

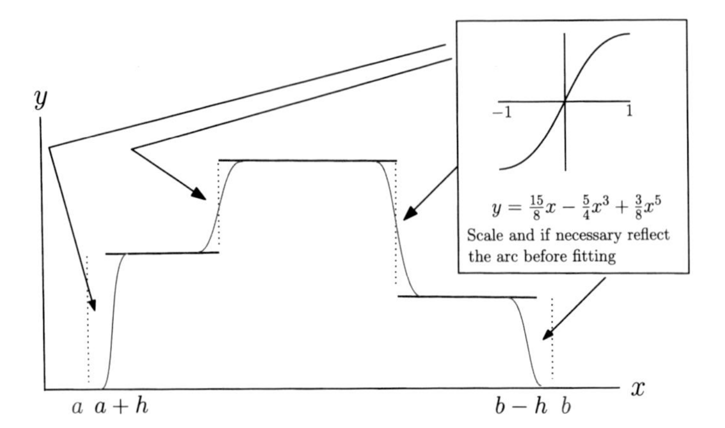

Fig. 7.1 Approximating a step function by a C^2 function in the mean

Finally, we approximate g in the mean square by linear combinations of the functions ϕ_n. By Proposition 7.6 we have

$$g = \sum_{n=1}^{\infty} \langle g, \phi_n \rangle \phi_n$$

with uniform convergence. Therefore, there exists N such that

$$\sup_{a \le x \le b} \left| g(x) - \sum_{n=1}^{N} \langle g, \phi_n \rangle \phi_n(x) \right| < \varepsilon/3(b-a)^{1/2}.$$

But then we have

$$\left\| g - \sum_{n=1}^{N} \langle g, \phi_n \rangle \phi_n \right\|_2 < \varepsilon/3.$$

Finally, therefore

$$\left\| f - \sum_{n=1}^{N} \langle g, \phi_n \rangle \phi_n \right\|_2 < \|f - g\|_2 + \left\| g - \sum_{n=1}^{N} \langle g, \phi_n \rangle \phi_n \right\|_2 < \frac{2\varepsilon}{3} + \frac{\varepsilon}{3} = \varepsilon.$$

2. The formula is Parseval's equation, which we have seen is equivalent to the result stated in part 1. It is also a special case of the formula in part 3, itself often called Parseval's equation.

3. We have

$$\langle f, g\rangle - \sum_{n=1}^{N} \langle f, \phi_n\rangle\langle \phi_n, g\rangle = \Big\langle f - \sum_{n=1}^{N} \langle f, \phi_n\rangle \phi_n,\ g\Big\rangle.$$

By the Cauchy-Schwarz inequality the modulus of the right-hand is bounded by

$$\Big\| f - \sum_{n=1}^{N} \langle f, \phi_n\rangle \phi_n \Big\|_2 \|g\|_2$$

which tends to zero as $N \to \infty$ by item 1. □

7.2.3 Eigenvalue Problems with Weights

The solution of Laplace's equation in a circle or in a sphere throws up eigenvalue problems of the form

$$(p(x)y')' + q(x)y = \lambda r(x)y, \quad (a < x < b)$$

with homogeneous boundary conditions. The function $r(x)$ is here called a *weight function*. We assume it is real valued, strictly positive and continuous in the interval $[a, b]$. We set out some of the relevant facts, leaving the reader to verify them.

This problem can be handled by defining the operator

$$Ly := \frac{1}{r(x)}\Big((p(x)y')' + q(x)y\Big)$$

Now we have the ordinary eigenvalue problem

$$Ly = \lambda y$$

with homogeneous boundary conditions. However, we have to use an *inner product with weight function* $r(x)$, namely:

$$\langle u, v\rangle_w := \int_a^b r u \overline{v}.$$

The theory then proceeds exactly as for the problem with weight function 1, provided $p(x)$ and $q(x)$ are continuous in the closed interval $[a, b]$, and $p(x)$ has no zeros. Typical boundary conditions that make the problem self-adjoint (but now understood with respect to the inner product with weight function $r(x)$) are, as before, separated boundary conditions or periodic boundary conditions (the latter require that $p(a) = p(b)$). As in the case of the problem with weight function 1, a maximal sequence $(\phi_k)_{k=1}^{\infty}$ of normalised eigenfunctions is complete. Normalised is understood to mean

$$\langle \phi_k, \phi_k \rangle_w = 1.$$

The eigenfunction development of an integrable function f takes the form

$$f = \sum_{k=1}^{\infty} c_k \phi_k,$$

with mean square convergence in general, using the coefficients

$$c_k = \langle f, \phi_k \rangle_w = \int_a^b r(x) f(x) \overline{\phi_k(x)}\, dx.$$

Mean square convergence must be understood with the weight function $r(x)$, and means here that

$$\lim_{n\to\infty} \int_a^b r(x) \Big(f(x) - \sum_{k=1}^{n} c_k \phi_k(x) \Big)^2 dx = 0.$$

7.2.4 Exercises

1. Derive the most important examples of eigenfunction expansions by solving the following eigenvalue problems, all involving the same operator, but with various boundary conditions and on various intervals. In each case obtain the formula for the coefficient c_n in the eigenfunction expansion $f(x) = \sum c_n \phi_n(x)$. A worked example, essentially solving most of item (a) but on the interval [0, 1], can be found near the beginning of Sect. 1.5. An important point to bear in mind in all four items is that the obvious solution basis comprising $\cos\sqrt{\lambda}x$ and $\sin\sqrt{\lambda}x$ is not valid for $\lambda = 0$.

   (a) (Fourier sine series)

$$-y'' = \lambda y, \quad (0 < x < \pi),$$
$$y(0) = y(\pi) = 0.$$

(b) (Fourier cosine series)

$$-y'' = \lambda y, \quad (0 < x < \pi),$$
$$y'(0) = y'(\pi) = 0.$$

(c) (Fourier series; trigonometric form)

$$-y'' = \lambda y, \quad (-\pi < x < \pi),$$
$$y(-\pi) = y(\pi), \quad y'(-\pi) = y'(\pi).$$

(d) (Quarter range expansion)

$$-y'' = \lambda y, \quad (0 < x < \pi/2),$$
$$y(0) = 0, \quad y'(\pi/2) = 0.$$

2. Solve the following Sturm-Liouville eigenvalue problems, that is find all eigenvalues and eigenfunctions (the second has a weight function):

   (a) $-(x^2 y')' = \lambda y, \quad y(1) = y(2) = 0$
   (b) $-y'' = x^{-2}\lambda y, \quad y'(1) = y'(2) = 0.$

3. Analysis texts that cover Parseval's equation usually go to town with interesting applications to summing series, otherwise hard to decipher. Determine the sums of the following series (hints for a suitable eigenfunction expansion are given):

   (a) $\displaystyle\sum_{n=1}^{\infty} \frac{1}{(2n-1)^2}, \quad (f(x) = 1$, Fourier sine expansion on $[0, \pi])$
   (b) $\displaystyle\sum_{n=1}^{\infty} \frac{1}{n^4}, \quad (f(x) = x^2$, Fourier cosine expansion on $[0, \pi])$
   (c) $\displaystyle\sum_{n=1}^{\infty} \frac{1}{n^6}, \quad (f(x) = x(x^2 - \pi^2)$, Fourier sine expansion on $[0, \pi])$

   Note Items (b) and (c) are values of the Riemann zeta function, $\zeta(s) = \sum_{n=1}^{\infty} n^{-s}$, namely $\zeta(4)$ and $\zeta(6)$, whilst (a) gives (after a brief calculation) the value $\zeta(2) = \pi^2/6$ (the elucidation of which is called the Basel problem). Euler obtained a general formula relating the zeta function at even integers to the Bernoulli numbers B_{2m}, namely:

$$\zeta(2m) = (-1)^{m+1} \frac{(2\pi)^{2m} B_{2m}}{2(2m)!}$$

   No comparable formula is known for $\zeta(2m-1)$.

4. We study the problem

$$-y'' = \lambda y, \quad (0 < x < L)$$

$$\alpha y(0) - y'(0) = 0, \quad \beta y(L) + y'(L) = 0.$$

The constants α and β are real and *not both zero*. The boundary conditions, so-called Robin conditions, have a practical significance in the theory of heat conduction. In the following, the quantity μ is real and positive.

(a) Show that μ^2 is an eigenvalue if and only if

$$\tan \mu L = \frac{\mu(\alpha + \beta)}{\mu^2 - \alpha\beta}.$$

If μ is a root of this equation, write down a formula for the eigenfunction. *Note* We must count μ as a root if $\mu^2 = \alpha\beta$ and $\cos \mu L = 0$.

(b) Show that 0 is an eigenvalue if and only if $\alpha\beta \neq 0$ and

$$L = -\frac{\alpha + \beta}{\alpha\beta}.$$

What is the eigenfunction in these cases?

(c) Show that $-\mu^2$ is an eigenvalue if and only if

$$\tanh \mu L = -\frac{\mu(\alpha + \beta)}{\mu^2 + \alpha\beta}.$$

If μ is root of this equation, write down a formula for the eigenfunction.

(d) Suppose that $\beta = 0$. Show that if $\alpha > 0$ there is no negative eigenvalue, but if $\alpha < 0$ there is exactly one.
Note The maximum number of negative eigenvalues is in general two for Robin conditions. See Exercise B9.

(e) Suppose that $\beta = 0$. Let the *positive* eigenvalues be enumerated as an increasing sequence $\lambda_1, \lambda_2, \lambda_3, \ldots$. Show that if $\alpha > 0$ then $|\lambda_n - (n-1)\pi| \to 0$, whilst if $\alpha < 0$ then $|\lambda_n - n\pi| \to 0$.

5. Consider the regular Sturm-Liouville operator $Ly := (py')' + qy$ on the interval $[a, b]$, suppose that $p < 0$ and let $Q = \min q$. Recall from Sect. 7.1 Exercise 3 that the Rayleigh quotient satisfies $R(u) \geq Q$ in the following self-adjoint cases:

(i) The boundary conditions are separated, that is,

$$\alpha_1 y(a) + \alpha_2 y'(a) = 0, \quad \beta_1 y(b) + \beta_2 y'(b) = 0,$$

with $\alpha_1\alpha_2 \leq 0$ and $\beta_1\beta_2 \geq 0$.

(ii) $p(a) = p(b)$ and the boundary conditions are periodic.

Show that in both cases the eigenvalues lie in the interval $[Q, \infty[$ and are therefore bounded below.

6. Consider the regular Sturm-Liouville operator $Ly := (py')' + qy$ on the interval $[a, b]$, equipped with homogeneous boundary conditions for which the operator is self-adjoint. Let $(\phi_n)_{n=1}^{\infty}$ be a maximal orthonormal sequence of eigenfunctions, and let λ_n be the eigenvalue to which ϕ_n belongs. Let $R(u)$ be the Rayleigh quotient. Derive the formula

$$R(u) = \frac{\sum_{n=1}^{\infty} \lambda_n |\langle u, \phi_n\rangle|^2}{\sum_{n=1}^{\infty} |\langle u, \phi_n\rangle|^2}$$

for functions $u \in C^2[a, b]$, $(u \neq 0)$, that satisfy the boundary conditions.

7. In the context of the previous exercise assume that the eigenvalues are bounded below. Let the items be arranged so that the sequence of eigenvalues $(\lambda_n)_{n=1}^{\infty}$ is increasing. Note that the sequence $(\lambda_n)_{n=1}^{\infty}$ does not necessarily increase strictly; a double eigenvalue will appear twice, in successive places.

   (a) Show that

   $$\lambda_1 = \min\langle Lu, u\rangle$$

   where the minimum is taken over all functions $u \in C^2[a, b]$, that satisfy the boundary conditions $U_1(u) = U_2(u) = 0$ and the normalisation condition $\|u\|_2 = 1$.

   Note The meaning is that λ_1 is not just the infimum: it is attained and is therefore the minimum. This answers the second question about the Rayleigh quotient posed in Sect. 7.1.

   (b) Show that

   $$\lambda_n = \min_{u \in F_n,\ \|u\|_2 = 1} \langle Lu, u\rangle$$

   where F_n is the subspace of $C^2[a, b]$ consisting of functions that satisfy the boundary conditions and are orthogonal to $\phi_1, \ldots, \phi_{n-1}$.

   (c) (Min-max principle) Show that

   $$\lambda_n = \min_{W} \max_{u \in W,\ \|u\|_2 = 1} \langle Lu, u\rangle$$

   where the minimum is taken over all n-dimensional subspaces W of $C^2[a, b]$, comprising functions that satisfy the boundary conditions.

Hint for (c) An n-dimensional subspace of $C^2[a, b]$ must contain a non-zero function orthogonal to $\phi_1, \ldots, \phi_{n-1}$.

Note In the subsequent project B the question as to when the eigenvalues are bounded below will taken up again. It will turn out that the eigenvalues are always bounded below if $p < 0$ and the boundary conditions separated. The restrictions of Sect. 7.1 Exercise 3 for the coefficients in separated boundary conditions are therefore unnecessary.

8. Determine the minimum of the integral $\int_0^1 |y'|^2$ over all functions in $C^2[0, 1]$ that satisfy $\int_0^1 |y|^2 = 1$ and the boundary conditions defined in the following cases:

   (a) $y(0) = y(1) = 0$
   (b) $y'(0) = y'(1) = 0$
   (c) $y(0) = y'(1) = 0$

   Hint For each case, interpret $\int_0^1 |y'|^2$ as a Rayleigh quotient.
   Note The minima turn out to be the same if $y(x)$ is taken to be in the larger space $C^1[0, 1]$ or even if it is only piece-wise C^1.

9. Show that the lowest eigenvalue λ_1 of the problem

$$-y'' + x^2 y = \lambda y, \quad (0 < x < 1)$$
$$y(0) = 0, \quad y(1) = 0.$$

   satisfies $\pi^2 < \lambda_1 < 10\frac{2}{7}$.
   Hint Compute the Rayleigh quotient for a well chosen test function to obtain an upper bound. The bound $10\frac{2}{7}$ results from a polynomial test function. It is easy to do even better by using a circular function. To obtain a lower bound compare the Rayleigh quotient of this problem with that of the problem $-y'' = \lambda y$. The next exercise extends this idea.

10. We compare the eigenvalues of two regular Sturm-Liouville problems, on the same interval $[a, b]$ and with the same boundary conditions. The operators are

$$Ly := (p(x)y')' + q(x)y, \quad \tilde{L}y := (\tilde{p}(x)y')' + \tilde{q}(x)y$$

   where $p < 0$ and $\tilde{p} < 0$. Assume the boundary conditions

$$y(a) = 0, \quad y(b) = 0.$$

   Let $(\lambda_n)_{n=1}^{\infty}$ and $(\tilde{\lambda}_n)_{n=1}^{\infty}$ denote the eigenvalues of the two operators. Since they are known to be bounded below we may assume them arranged in increasing order.

   (a) Assume that $\tilde{p}(x) \leq p(x)$, $\tilde{q}(x) \geq q(x)$ for all x in $[a, b]$. Show that $\lambda_n \leq \tilde{\lambda}_n$ for $n = 1, 2, 3, ...$
   (b) Obtain the same result for periodic boundary conditions (given that $p(a) = p(b)$). In this case a double eigenvalue is possible and it is assumed that such eigenvalues appear twice, in successive places, in the sequence.

Hint Compare the Rayleigh quotients and use the min-max principle, Exercise 7(c).

11. We consider a regular Sturm-Liouville operator L on an interval $[a, b]$, self-adjoint with given boundary conditions. Let $G(x, \xi, \lambda)$ be the Green's function for $L - \lambda$. Let a maximal orthonormal set of eigenfunctions be enumerated as the sequence $(\phi_n)_{n=1}^{\infty}$ with eigenvalue λ_n corresponding to ϕ_n.

    (a) Show that if λ is not an eigenvalue then

    $$\phi_n(x) = (\lambda_n - \lambda) \int_a^b G(x, \xi, \lambda)\phi_n(\xi)\, d\xi.$$

    (b) Deduce that the expansion

    $$G(x, \xi, \lambda) = \sum_{n=1}^{\infty} \frac{\phi_n(x)\, \overline{\phi_n(\xi)}}{\lambda_n - \lambda}$$

    holds, at least in the weak sense that for fixed x and λ it converges in the mean square for $a < \xi < b$.
    Note It is not hard to show, as suggested by the series, that $-\phi_n(x)\, \overline{\phi_n(\xi)}$ is the residue of $G(x, \xi, \lambda)$ at $\lambda = \lambda_n$. Thus, if the eigenvalues are known, and are known to be simple, the eigenfunctions, fully normalised, can be obtained in a rather direct fashion from the Green's function. This is an interesting payoff, since given unnormalised eigenfunctions it may be tricky to calculate the normalisation factors.

    (c) Show that

    $$\overline{G(x, \xi, \lambda)} = G(\xi, x, \overline{\lambda}).$$

    (d) Show that

    $$\int_a^b |G(x, \xi, \lambda)|^2\, d\xi = \sum_{n=1}^{\infty} \frac{|\phi_n(x)|^2}{|\lambda - \lambda_n|^2}.$$

    (e) Obtain the formula

    $$\int_a^b \int_a^b |G(x, \xi, \lambda)|^2\, d\xi\, dx = \sum_{n=1}^{\infty} \frac{1}{|\lambda - \lambda_n|^2}.$$

    Note If you are using the Riemann integral you would probably want to justify the term-by-term integration by trying to prove that the series is uniformly convergent for $a < x < b$. This is quite tricky. The result follows at once if we think of the integrals as Lebesgue integrals and use the monotone convergence theorem.

(f) Deduce that the series

$$\sum_{\lambda_n \neq 0} \frac{1}{|\lambda_n|^2}$$

is convergent.

(g) In the case of an eigenvalue problem with weight function $r(x)$ show that the correct series is

$$r(\xi)^{-1} G(x, \xi, \lambda) = \sum_{n=1}^{\infty} \frac{\phi_n(x)\,\overline{\phi_n(\xi)}}{\lambda_n - \lambda}$$

7.2.5 *Projects*

A. *Project on classical Fourier series*

In this project we show how ideas introduced in this chapter can be used to obtain proofs of some classical results of Fourier series. In particular, we obtain results on uniform convergence and mean square convergence. Of course, the many treatises that focus exclusively on these series present conclusions that go much further.

We study the *first order self-adjoint eigenvalue problem*

$$\frac{1}{i} y' = \lambda y, \quad y(0) = y(2\pi).$$

Let L be the first order differential operator

$$Ly := \frac{1}{i} y'$$

acting on functions in the interval $]0, 2\pi[$. By "the boundary conditions" we mean the condition $y(0) = y(2\pi)$ (although there is only one). As usual, we use the inner product

$$\langle u, v \rangle := \int_0^{2\pi} u\overline{v}$$

for functions u and v integrable in $[0, 2\pi]$.

A1. (a) Show that the problem is self-adjoint, meaning that $\langle u, Lv \rangle = \langle Lu, v \rangle$ for all functions u and v in $C^1[a, b]$ that satisfy the boundary conditions.

(b) Find all eigenvalues and eigenfunctions of L with the boundary conditions.

Answer. The normalised eigenfunctions and eigenvalues may be written as two-sided sequences indexed by the integers (including the negative ones)

$$\phi_n(x) = \frac{1}{\sqrt{2\pi}} e^{inx}, \quad \lambda_n = n, \quad (n \in \mathbb{Z})$$

(c) Show that $\langle \phi_m, \phi_n \rangle = 0$ if $m \neq n$.

(d) Calculate the Green's function $G(x, \xi, \lambda)$ for the problem

$$Ly - \lambda y = h(x), \quad (0 < x < 2\pi), \quad y(0) = y(2\pi)$$

given that $\lambda \notin \mathbb{Z}$. Show explicitly that G has a simple pole at $\lambda = n$, for each integer n (positive, negative or 0), and calculate its principal part.

(e) Using the previous item show that if $f \in C[0, 2\pi]$ then the function

$$F(x, \lambda) := \int_0^{2\pi} G(x, \xi, \lambda) f(\xi)\, d\xi$$

is an entire analytic function of λ if and only if $\langle f, \phi_n \rangle = 0$ for all $n \in \mathbb{Z}$.

(f) Copying the proof of Proposition 7.2 (conclusion 3), and using the previous item, show that if $f \in C[0, 2\pi]$ and $\langle f, \phi_n \rangle = 0$ for all $n \in \mathbb{Z}$, then $f = 0$.

Now we may derive some classical theorems of Fourier series, by imitating the proofs of Propositions 7.6 and 7.10.

A2. (a) Let $f \in C^1[0, 2\pi]$ and suppose that $f(0) = f(2\pi)$. Then

$$f = \sum_{n=-\infty}^{\infty} \langle f, \phi_n \rangle \phi_n$$

with *absolute and uniform* convergence.

Note The same conclusion holds under the weaker assumption that f is continuous and f' piece-wise continuous (though we still need $f(0) = f(2\pi)$). This is because integration by parts, needed to verify that $\langle f, L\phi_n \rangle = \langle Lf, \phi_n \rangle$, is still valid.

(b) Let f be integrable on the interval $[0, 2\pi]$. Then

$$f = \sum_{n=-\infty}^{\infty} \langle f, \phi_n \rangle \phi_n$$

in the mean square, in the strong sense that for all $\varepsilon > 0$ there exists a finite subset $A \subset \mathbb{Z}$, such that

$$\Big\| \sum_{n \in B} \langle f, \phi_n \rangle \phi_n - f \Big\|_2 < \varepsilon.$$

for all finite subsets $B \subset \mathbb{Z}$ that include A.

B. *Project on Prüfer's transformation*

The conclusion of Exercise 5, that the eigenvalues are bounded below if $p < 0$ and with specified boundary conditions, obtains quite generally if the boundary conditions are separated (and not just for the special separated boundary conditions of Exercise 5(a)). This result can be proved by returning to the idea suggested at the very beginning of this chapter. We set

$$u = y, \quad v = p(x)y' \tag{7.11}$$

and track the progress of the point (u, v) in a phase plane, given that y satisfies $(py')' + (q - \lambda)y = 0$ in the interval $a \le x \le b$. The change of variables (7.11) is called Prüfer's transformation. It can be used to obtain quite detailed information about the eigenvalues and eigenfunctions of a Sturm-Liouville problem. We emphasise that throughout this project we assume that $p < 0$ in $[a, b]$.

We consider the separated boundary conditions

$$\alpha_1 y(a) + \alpha_2 y'(a) = 0, \quad \beta_1 y(b) + \beta_2 y'(b) = 0$$

with the usual proviso that neither (α_1, α_2) nor (β_1, β_2) is the zero vector. Interpreted in the phase plane, the boundary conditions say that (u, v) starts when $x = a$ on the line ℓ_1, that has the equation

$$\alpha_1 u + (\alpha_2/p(a))v = 0,$$

and ends when $x = b$ on the line ℓ_2, with equation

$$\beta_1 u + (\beta_2/p(b))v = 0.$$

The number λ is an eigenvalue (we know in advance that the eigenvalues are real so we take λ to be real) if and only if a solution exists, starting on ℓ_1 and ending on ℓ_2, that does not pass through the origin $(0, 0)$. This is illustrated in Fig. 7.2.

Given that the phase curve does not pass through the origin, we can define a continuous phase angle $\omega(x)$, such that

$$u(x) + iv(x) = \big(u(x)^2 + v(x)^2\big)^{1/2} e^{i\omega(x)}, \quad (a \le x \le b).$$

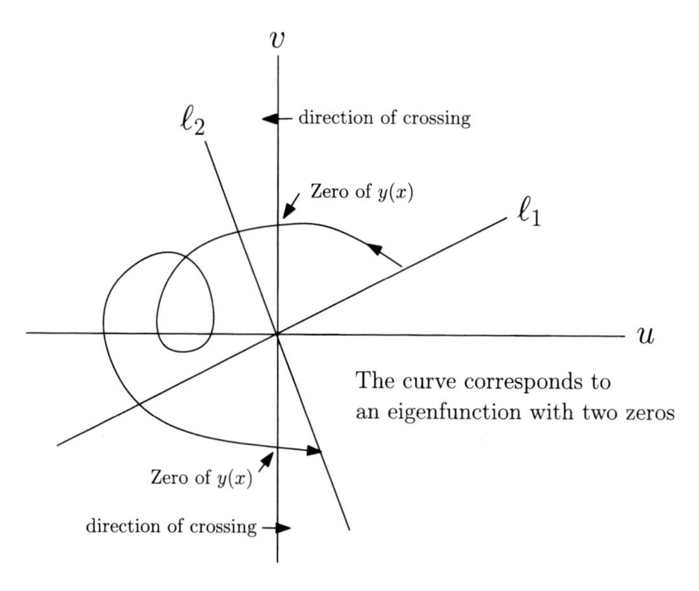

Fig. 7.2 Prüfer's transformation

We write

$$\omega \in [s, t], \quad (\text{mod } \pi)$$

to mean that $s \leq \omega + \pi m \leq t$ for some integer m.

B1. Show that $\omega(x)$ satisfies the differential equation

$$\frac{d\omega}{dx} = -\frac{1}{p(x)} \sin^2 \omega + \big(\lambda - q(x)\big) \cos^2 \omega. \tag{7.12}$$

B2. Show that $d\omega/dx > 0$ whenever $u = 0$. In other words the phase point crosses the v-axis in the direction of increasing phase angle.

B3. Let $-\pi/2 < \theta_1 \leq \theta_2 < \pi/2$ and let $L \geq 0$. Show that there exists $K > 0$ such that if $\lambda < -K$, then $d\omega/dx < -L$ for $a \leq x \leq b$, so long as $\omega(x) \in [\theta_1, \theta_2]$, (mod π). In other words, for sufficiently low λ we can make ω decrease arbitrarily fast with increasing x, as long as ω is between θ_1 and θ_2 up to an integer multiple of π.

B4. Show that the solution $\omega(x, \lambda)$ of (7.12) that satisfies an initial condition $\omega(a) = \theta_0$ is a strictly increasing function of λ for the range $a < x \leq b$. *Hint* Look at the equation of variation (see Sect. 6.2.3) satisfied by $\partial\omega/\partial\lambda$. It reduces to applying the very first result in this text, Proposition 1.1.

The first objective is to show that the eigenvalues are bounded below. We first consider boundary conditions such that neither α_2 nor β_2 is zero. This means that (u, v) neither starts nor ends on the v-axis. We can assign direction angles θ_1 to the line ℓ_1 and θ_2 to the line ℓ_2, such that θ_1 and θ_2 lie in the open interval $]-\pi/2, \pi/2[$. Now λ is an eigenvalue if the solution of (7.12) with the initial condition $\omega(a) = \theta_1$ satisfies $\omega(b) = \theta_2 + n\pi$ for some integer n. This is illustrated in Fig. 7.2. We don't need to consider the initial condition $\omega(a) = \theta_1 + \pi$ because if $\omega(x)$ is a solution of (7.12) so is $\omega(x) - \pi$.

B5. In the case that $\theta_1 \le \theta_2$, show that no number below $-K$, where K is defined in Exercise B3 by taking $L = 0$, is an eigenvalue.

B6. In the case that $\theta_1 > \theta_2$ (as drawn in Fig. 7.2), show that there exists $k > 0$, such that if $\lambda < -k$ then the solution that satisfies $\omega(a) = \theta_1$ must satisfy $-\pi/2 < \omega(b) < \theta_2$, so that λ cannot be an eigenvalue.

In the remaining cases either $\alpha_2 = 0$ or $\beta_2 = 0$, so that (u, v) begins, or ends, on the v-axis, or both. Some of these cases were covered in Exercise 5, but they can all be handled by the ideas set forth here.

B7. Find arguments to show that the eigenvalues are bounded below in the remaining cases.

A second objective is to obtain useful information about the zeros of eigenfunctions, given that we already know that the eigenvalues form an infinite sequence tending to infinity.

B8. Let the eigenvalues be arranged in ascending order $\lambda_1 < \lambda_2 < \cdots$ with eigenfunction ϕ_k belonging to λ_k. Prove that ϕ_k has $k - 1$ zeros in the open interval $]a, b[$.
Hint The solution $y(x)$ has a zero when the point (u, v) crosses the v-axis (Fig. 7.2). Begin with the situation as described in Exercises B5 and B6, and use the fact that $\omega(b, \lambda)$ is strictly increasing with λ (Exercise B4).

B9. The result of the previous exercise can be used to clarify the problem $-y'' = \lambda y$ with Robin boundary conditions (Exercise 4)

$$\alpha y(0) - y'(0) = 0, \quad \beta y(L) + y'(L) = 0.$$

(a) Show that there cannot be more than two negative eigenvalues.
Hint Show that the lowest positive eigenvalue is below $4\pi^2$ and therefore its eigenfunction has at most 2 zeros.

(b) Show that if $L = 1$, $\alpha < 0$, $\beta < 0$, and $|\alpha| + |\beta| < \alpha\beta$, then there are exactly two negative eigenvalues.
Hint Let $\lambda = -\mu^2$, with $\mu > 0$. By Exercise 4, λ is an eigenvalue if and only if μ is a (positive) root of

$$\tanh \mu = \frac{\mu(|\alpha| + |\beta|)}{\mu^2 + \alpha\beta}.$$

Show that for $\mu > 0$ the right-hand side attains a maximum above 1 and deduce that there are at least two roots.[8]

B10. Without assuming the existence of eigenvalues, give a new proof that the eigenvalues form an infinite sequence unbounded above, by showing, on the basis of (7.12), that $\lim_{\lambda\to\infty} \omega(b,\lambda) = \infty$.
Hint Let

$$A = \min_{a\le x\le b}\left(-\frac{1}{p(x)}\right), \quad B = \max_{a\le x\le b} q(x).$$

Then

$$\frac{d\omega}{dx} \ge A\sin^2\omega + (\lambda - B)\cos^2\omega$$

and if $\lambda > B$ we find (separating the variables and letting θ_0 be the initial value for ω):

$$\int_{\theta_0}^{\omega(b,\lambda)} \frac{d\omega}{A\sin^2\omega + (\lambda - B)\cos^2\omega} \ge b - a.$$

If $\lim_{\lambda\to\infty} \omega(b,\lambda) = \Omega < \infty$, a contradiction results from letting λ tend to infinity.

C. *Project on singular Sturm-Liouville problems*

One of the principal applications of Sturm-Liouville theory is to the solution of partial differential equations by the method of separation of variables. In the course of this the student quickly encounters Sturm-Liouville problems that are not regular. Typically this occurs because the function $p(x)$ is zero at one or both endpoints. For these *singular* Sturm-Liouville problems, extending the notion of eigenfunction expansion is a major challenge and proves quite technical. In this project we shall study two important examples.

First we return to the Legendre equation, already studied in 3.1 project A. We define the operator

$$Ly := ((x^2 - 1)y')'$$

and pose the eigenvalue problem

$$Ly = \lambda y, \quad (-1 < x < 1).$$

[8] Some books assert that is it clear from a sketch of the graphs alone that there are *exactly two roots*. This is far from obvious.

We still have to specify boundary conditions. However, even before we turn to that question, there are some problems that must be faced. In the case of the regular Sturm-Liouville problem, we worked with the function space $C^2[a, b]$ as a natural domain for the differential operator. Functions in this space, together with their first and second derivatives, have boundary values at a and b. However, the equation $Ly - \lambda y = 0$ has singular points at -1 and 1, and we cannot solve it with arbitrarily assigned Cauchy data at an endpoint; in fact it has solutions that blow up at an endpoint. Therefore, as a putative domain for the operator L the space $C^2[-1, 1]$ is too restrictive; we must work with functions that do not necessarily have boundary values. We shall assume, for a start, that the operator L *is applied to functions that have a continuous second order derivative in the open interval* $]-1, 1[$. We don't require boundary values to exist.

However, there is another problem. We want the integrals $\int_a^b u\overline{v}$ and $\int_a^b u\overline{Lv}$ to exist, at least as absolutely convergent integrals. So we need to impose a further condition on the domain of the operator L. We shall assume that L is applied *only to functions* $y(x)$ *that satisfy*

$$\int_{-1}^{1} |y(x)|^2\,dx < \infty, \quad \int_{-1}^{1} |Ly(x)|^2\,dx < \infty$$

where the integrals are absolutely convergent. It is left to the reader to show that if u and v satisfy these conditions then the integral $\int_{-1}^{1} u\overline{Lv}$ is convergent.

With this understanding about the *domain of* L, we seek linear side conditions, that are homogeneous (that is, they are satisfied by the zero function) and give rise to the symmetry condition. More precisely, if u and v are in the domain of L and satisfy the side conditions, then we require

$$\langle Lu, v\rangle = \langle u, Lv\rangle,$$

where, as usual, the inner product is given by

$$\langle u, v\rangle = \int_{-1}^{1} u\overline{v}.$$

C1. Show that if u and v are in the domain of L, then

$$\langle Lu, v\rangle - \langle u, Lv\rangle = \lim_{x\to 1-} (x^2 - 1)(u'(x)\overline{v(x)} - u(x)\overline{v'(x)}) - \lim_{x\to -1+} (x^2 - 1)(u'(x)\overline{v(x)} - u(x)\overline{v'(x)}).$$

We take a naive approach, guided by the formula in the previous item, and impose "boundary conditions":

$$(\mathrm{B}_1): \quad y(x) \text{ is bounded as } x \to -1, \qquad \lim_{x \to -1+} (1+x)y'(x) = 0$$

$$(\mathrm{B}_2): \quad y(x) \text{ is bounded as } x \to 1, \qquad \lim_{x \to 1-} (1-x)y'(x) = 0$$

C2. (a) Show that if u and v are in the domain of L and satisfy the boundary conditions B_1 and B_2, then

$$\langle Lu, v\rangle = \langle u, Lv\rangle.$$

(b) Consider the eigenvalue problem $Ly = \lambda y$ with the boundary conditions B_1 and B_2. Show that eigenvalues (if any exist) are real, and that eigenfunctions belonging to distinct eigenvalues are orthogonal to each other.

Of course we know that eigenvalues exist. We saw in Sect. 3.1 project A (Exercise 3.1.4) that a non-trivial solution, bounded in $]-1, 1[$, exists if and only if λ is of the form $l(l+1)$, where l is a natural number, and the solution is a polynomial of degree l. Up to a scalar multiple, it is the Legendre polynomial of degree l, denoted by $P_l(x)$. We form the orthonormal sequence

$$\phi_l(x) = \Lambda_l^{-1} P_l(x),$$

where the positive normalisation constants Λ_l satisfy

$$\Lambda_l^2 = \int_{-1}^{1} P_l(x)^2 \, dx.$$

C3. Show that

$$\Lambda_l = \sqrt{\frac{2}{2l+1}}.$$

Hint Use Rodrigues's formula (Sect. 3.1 exercise A8), and integrate repeatedly by parts.

We can give a direct proof of the completeness of the orthonormal sequence $(\phi_l(x))_{l=0}^{\infty}$.

C4. (a) Show that the sequence ϕ_0, ..., ϕ_N forms a basis for the vector space of polynomials with degree less than or equal to N.

(b) Prove that the orthonormal sequence $(\phi_l(x))_{l=0}^{\infty}$ is complete on the interval $[-1, 1]$.

Hint for item (b) Every function integrable on $[-1, 1]$ can be approximated in the mean square by continuous functions, and, by the Weierstrass approximation theorem, every function continuous in $[-1, 1]$ can be approximated uniformly by polynomials. It is therefore enough, by Proposition 7.9, to show that every polynomial f can be approximated in the mean square on $[-1, 1]$ by a linear combination of Legendre polynomials. To show this, use item (a).

The second example we consider here is the problem

$$-(xy')' + \frac{\alpha^2}{x}y = \lambda xy, \quad (0 < x < 1)$$

with the boundary condition $y(1) = 0$, and a boundary condition at 0 still to be specified. The constant α is real and non-negative. This problem arises from solving the wave equation in a circular disc.

Here the singular features are two: the function $p(x) := -x$ is zero at the endpoint $x = 0$; and the function $q(x) := \alpha^2/x$ is singular at the endpoint $x = 0$. In addition we have a weight function $r(x) := x$. This we handle by defining

$$Ly := \frac{1}{x}\left(-(xy')' + \frac{\alpha^2}{x}y\right)$$

and using the weighted inner product

$$\langle u, v\rangle_r := \int_0^1 xu(x)\overline{v(x)}\,dx.$$

For the domain of L we need in the first place functions that have a continuous second order derivative in $]0, 1[$, but don't necessarily have boundary values, and in addition, to ensure the existence of the necessary integrals we shall require that functions $y(x)$ in the domain of L satisfy

$$\int_0^1 x|y(x)|^2\,dx < \infty, \quad \int_0^1 x|Ly(x)|^2\,dx < \infty.$$

C5. Show that if u and v are are in the domain of L and satisfy $u(1) = v(1) = 0$ then:

$$\langle Lu, v\rangle_r - \langle u, Lv\rangle_r = \lim_{x\to 0+} x(u'(x)\overline{v(x)} - u(x)\overline{v'(x)})$$

We therefore define the boundary conditions:

$$(B_1): \quad y(x) \text{ is bounded as } x \to 0+, \qquad \lim_{x\to 0+} xy'(x) = 0$$

$$(B_2): \quad y(1) = 0.$$

C6. (a) Show that if u and v are in the domain of L and satisfy the boundary conditions B_1 and B_2, then

$$\langle Lu, v\rangle_r - \langle u, Lv\rangle_r = 0.$$

(b) Show that all eigenvalues (if any) of the problem

$$Ly = \lambda y, \quad (0 < x < 1), \quad \text{with } B_1 \text{ and } B_2,$$

are real, and that eigenfunctions belonging to distinct eigenvalues are orthogonal to each other.

(c) Show that

$$\langle Ly, y\rangle_r = \int_0^1 \left(x|y'(x)|^2 + \alpha^2 x^{-1}|y(x)|^2\right) dx \geq 0$$

and deduce that all eigenvalues are non-negative.

Next we find the eigenvalues and eigenfunctions. We rewrite the equation as

$$x^2 y'' + xy' + (\lambda x^2 - \alpha^2)y = 0.$$

C7. (a) Show that 0 is not an eigenvalue.

(b) Knowing that λ is real and positive, we set $t = \sqrt{\lambda}\,x$. Show that this leads to the problem:

$$t^2 \frac{d^2y}{dt^2} + t\frac{dy}{dt} + (t^2 - \alpha^2)y = 0.$$

with boundary conditions requiring $y = 0$ at $t = \sqrt{\lambda}$, whilst at 0 we require that y is bounded and $t\,dy/dt$ tends to 0.

(c) Show that the most general solution that satisfies the boundary condition at $t = 0$ is $AJ_\alpha(t)$, (where J_α is the Bessel function of the first kind of order α; see Sect. 3.2, project on the Bessel equation), so that in terms of x we have:

$$y(x) = AJ_\alpha(\sqrt{\lambda}x).$$

(d) Every Bessel function has infinitely many zeros (Sect. 3.2 Exercise A5). Let the zeros of J_α be arranged as an increasing sequence

$$\mu_1 < \mu_2 < \mu_3 \cdots \to \infty.$$

Show that the eigenvalues form the sequence

$$\mu_1^2 < \mu_2^2 < \mu_3^2 \cdots \to \infty$$

and a not yet normalised eigenfunction corresponding to μ_k^2 can be taken to be $J_\alpha(\mu_k x)$.

We have therefore found *all eigenvalues and the corresponding eigenfunctions* for our problem. Next we form the orthonormal sequence

$$\psi_k(x) = \Gamma_n^{-1} J_\alpha(\mu_k x), \quad (k = 1, 2, ...)$$

where the positive normalisation constants Γ_k satisfy

$$\Gamma_k^2 = \int_0^1 x J_\alpha(\mu_k x)^2 \, dx.$$

C8. Show that

$$\Gamma_k = \frac{1}{\sqrt{2}} |J_\alpha'(\mu_k)| = \frac{1}{\sqrt{2}} |J_{\alpha+1}(\mu_k)|.$$

Hint Use Sect. 3.2 Exercises A3 and A4.

It is a fact that the orthonormal sequence $(\psi_k(x))_{k=1}^\infty$ is complete on the interval $[0, 1]$. The corresponding expansion, called the Fourier-Bessel expansion, allows the development of an integrable function f in a series, in general convergent in the mean square:

$$f(x) = \sum_{k=1}^\infty c_k \psi_k(x), \quad (0 < x < 1),$$

where

$$c_k = \int_0^1 x f(x) \psi_k(x) \, dx, \quad (k = 1, 2, ...).$$

Mean square convergence must be understood with the weight function; in the present case:

$$\lim_{n \to \infty} \int_0^1 x \Big(f(x) - \sum_{k=1}^n c_k \psi_k(x) \Big)^2 dx = 0.$$

A direct proof of completeness along the lines that we used for the Legendre polynomials does not seem feasible. It, like other completeness results, are usually obtained by the construction and close examination of a Green's function for the problem $Ly - \lambda y = h(x)$, with the boundary conditions, which is how we proceeded in the case of regular Sturm-Liouville problems. The best way forward for singular problems is via the theory of self-adjoint operators in a Hilbert space. This requires the Lebesgue integral, and a more sophisticated approach to domains and boundary conditions than is possible in the present text. We shall accept these facts and consign them to the area of further studies. On this forward looking note, we bring this survey of essential ordinary differential equations to a natural close.

Afterword

It would be possible to compose a long bibliography of works on ordinary differential equations, many of them encyclopedic, others more like first courses. However, two things might be more useful to the reader: hints about where to obtain more information about topics already in the current text; and pointers to further studies. We list some books, mostly classics, and not in any logical order (though the two oldest come first), with short comments about how they may prove useful to the reader.

1. E. T. Whittaker and G. N. Watson. A Course of Modern Analysis.

First published in 1902 it remains an indispensable source to this day. The extensive coverage of non-elementary transcendental functions (often called special functions; Bessel functions, Legendre functions etc.), that arise as solutions of the differential equations of mathematical physics, provides an opportunity for further studies to readers of the current text.

2. E. L. Ince. Ordinary Differential Equations.

Originally published 1926 it is probably the oldest textbook on differential equations in English that is still widely read and consulted. A whole chapter is devoted to continuous transformation groups, a most unusual feature. The treatment of the Sturm-Liouville development of a function by complex analysis is the basis of the proofs given in the current text.

3. E. A. Coddington and N. Levinson. Theory of Ordinary Differential Equations.

An admirable and much used textbook. For the further studies of readers of the current text we can mention its final three chapters covering plane autonomous systems, the Poincaré-Bendixson theorem, and differential equations on a torus.

R. Magnus, *Essential Ordinary Differential Equations*, Springer Undergraduate Mathematics Series, https://doi.org/10.1007/978-3-031-11531-8

4. V. I. Arnold. Ordinary Differential Equations.

A delightful book that opened up new ways to approach the teaching of ordinary differential equations. The reader is often charmed by the author's characteristic geometrical way of viewing the subject.

5. G. Birkhoff and G-C. Rota. Ordinary Differential Equations.

A much used textbook covering all the basic material. It includes a chapter on numerical methods (a sin of omission of the current text).

6. W. Hurewicz. Lectures on Ordinary Differential Equations.

Overall a bit challenging. The chapter on plane autonomous systems, in particular the classification of equilibrium points, is well worth reading, under the heading of further studies.

7. W. A. Strauss. Partial Differential Equations.

The Sturm-Liouville theory of the current text is the basis of one of the principal methods for solving problems in partial differential equations: the method of separation of variables. Bessel's equation, the Legendre equation, separated boundary conditions, and more, all make their appearance in the solution of Laplace's equation, the heat equation and the wave equation in physically important domains.

8. A. E. Taylor. Introduction to Functional Analysis.

The theory of compact self-adjoint operators in a Hilbert space provides the clearest and most decisive way to justify eigenfunction expansions in the context of Sturm-Liouville problems, both regular and singular. For this the student must study functional analysis. This book has maintained its reputation since its first publication in 1958. It also includes excellent coverage of the operational calculus of a linear operator, which extends to infinite dimensional spaces the analytic functions of a matrix covered in Chap. 5 of the present text.

9. N. Levinson and R. M. Redheffer. Complex Variables.

A very readable introduction to complex analysis. It contains all the material on Laurent series needed for Chap. 7 of the present text.

10. G. B. Folland. Fourier Analysis and its Applications.

A most readable and informative text on Fourier analysis which also covers Sturm-Liouville theory and its applications to solving classical problems of partial differential equations.

Finally a word about the epigraphs to each chapter. The reader may have recognised that they are taken from James Joyce's Ulysses. This year marks one hundred years since the publication of that rich and complex web of cross references.

Index

R. Magnus, *Essential Ordinary Differential Equations*, Springer Undergraduate Mathematics Series, https://doi.org/10.1007/978-3-031-11531-8

Printed in the United States
by Baker & Taylor Publisher Services